Lecture Notes in Statistics

Edited by D. Brillinger, S. Fienberg, J. Gani,
J. Hartigan, J. Kiefer, and K. Krickeberg

W0246149

4

Erik van Doorn

Stochastic Monotonicity and Queueing Applications of Birth–Death Processes

Springer-Verlag
New York   Heidelberg   Berlin

E. A. van Doorn
Netherlands Postal and Telecommunications Services
Dr. Neher – Laboratories
Post Office Box 421
2260 AK Leidschendam
The Netherlands

This monograph is a polished version of the author's dissertation entitled:
"Stochastic Monotonicity of the Birth–Death Processes," which was written
while he was affiliated with the Department of Applied Mathematics,
Twente University of Technology, Enschede.

AMS Subject Classifications (1980): primary 60J80; secondary 60K25, 92A15

Library of Congress Cataloging in Publication Data

Doorn, Erik van.
   Stochastic monotonicity and queueing applications of birth-death processes.

   (Lecture notes in statistics; 4)
   Based on the author's thesis, Twente University
of Technology, Enschede.
   Bibliography: p.
   Includes indexes.
   1. Birth and death processes (Stochastic
processes) 2. Monotone operators. 3. Queueing
theory. I. Title. II. Series.
QA274.76.D66   519.2'34   80-25183

ISBN 978-0-387-90547-1   ISBN 978-1-4612-5883-4 (eBook)

DOI 10.1007/978-1-4612-5883-4

9 8 7 6 5 4 3 2 1

PREFACE

A stochastic process $\{X(t): 0 \le t < \infty\}$ with discrete state space $S \subset \mathbb{R}$ is said to be stochastically increasing (decreasing) on an interval T if the probabilities $Pr\{X(t) > i\}$, $i \in S$, are increasing (decreasing) with t on T. Stochastic monotonicity is a basic structural property for process behaviour. It gives rise to meaningful bounds for various quantities such as the moments of the process, and provides the mathematical groundwork for approximation algorithms.

Obviously, stochastic monotonicity becomes a more tractable subject for analysis if the processes under consideration are such that stochastic monotonicity on an interval $0 < t < \varepsilon$ implies stochastic monotonicity on the entire time axis. DALEY (1968) was the first to discuss a similar property in the context of discrete time Markov chains. Unfortunately, he called this property "stochastic monotonicity", it is more appropriate, however, to speak of processes with monotone transition operators. KEILSON and KESTER (1977) have demonstrated the prevalence of this phenomenon in discrete and continuous time Markov processes. They (and others) have also given a necessary and sufficient condition for a (temporally homogeneous) Markov process to have monotone transition operators. Whether or not such processes will be stochastically monotone as defined above, now depends on the initial state distribution. Conditions on this distribution for stochastic monotonicity on the entire time axis to prevail were given too by KEILSON and KESTER (1977).

It is very well conceivable that a process with monotone transition operators is not stochastically monotone on the entire positive time axis but on an interval of the form (t_1, ∞), with $t_1 > 0$. Clearly, it is of some interest to know under which circumstances this phenomenon occurs. The study of these circumstances is the main subject of this monograph. The analysis is restricted to birth-death processes, which form the most important class of temporally homogeneous Markov processes in continuous time with monotone transition operators. In some proofs explicit use is made of specific properties of birth-death processes, so that it is probably not possible to extend the results to other classes of Markov processes.

The main results of this monograph are obtained in chapter 5 where necessary and sufficient conditions are given for a birth-death process to be stochastically monotone in the long run when the state space is a semi-infinite lattice of integers and the initial state distribution is supported by finitely many points. The theory needed to arrive at these results is developed in the chapters 3 and 4. It appears that the concept of dual processes, which is only touched upon in the existing literature, is very fruitful and of intrinsic interest.

In chapter 1 and chapter 2 many known facts about birth-death processes are collected. Also in chapter 2 some preliminary analysis is done with a view to the chapters 6, 7 and 8, where the results are applied to specific processes. To do this one needs iv

at least partial knowledge of the so-called spectral representation of the transition
probabilities. As for the linear growth, birth-death processes of chapter 8 (including
the M/M/∞ queue length process) this knowledge is available and application of the
results of chapter 5 to these processes is straightforward. This is not the case,
however, with the M/M/s queue length process of chapter 6 and the queue length process
of chapter 7 which models a system where potential customers are discouraged by queue
length. A substantial part of this monograph, in fact the main part of the chapters 6
and 7 is therefore concerned with obtaining these representations, which have an
interest of their own. Our findings in this respect extend the results previously
obtained by KARLIN and McGREGOR (1958a) and NATVIG (1974), respectively.
In chapter 9 various aspects of the first moment of birth-death processes are discussed. It appears to behave very regularly in a number of important cases.
Finally, birth-death processes with a finite state space are considered in chapter 10.
Although the analysis of the phenomenon of stochastic monotonicity may be performed
through the concept of dual processes as in the infinite case, an entirely different
approach is chosen.

I take pleasure in closing this preface by acknowledging the support of Professor
Jos H.A. de Smit of Twente University of Technology who provided the key references,
and by thanking Miss Bea Bhola of the Dr. Neher - Laboratories for the careful typing
of the manuscript.

Leidschendam, August 1980   Erik van Doorn

TABLE OF CONTENTS

1.   PRELIMINARIES

1.1   Markov processes (CHUNG (1967), FREEDMAN (1971), REUTER (1957))

By a *Markov process* we shall understand a continuous time stochastic process
$\{X(t): 0 \leq t < \infty\}$ which has a denumerable state space S and which possesses the
Markov property, i.e., for every $n \geq 2$, $0 \leq t_1 < \ldots < t_n$ and any $i_1, \ldots, i_n$ in
S one has

(1.1.1) $\Pr\{X(t_n) = i_n \mid X(t_1) = i_1, \ldots, X(t_{n-1}) = i_{n-1}\} =$

$$\Pr\{X(t_n) = i_n \mid X(t_{n-1}) = i_{n-1}\},$$

The process is supposed to be *temporally homogeneous*, i.e., for every i, j in S
the conditional probability $\Pr\{X(t+s) = j \mid X(s) = i\}$ does not depend on s. In this
case we may put

(1.1.2) $p_{ij}(t) = \Pr\{X(t+s) = j \mid X(s) = i\}$ $t \geq 0.$

The function $p_{ij}(.)$ is the *transition probability function from the state i to
the state j*. The *absolute distribution at time* $t \geq 0$ is defined to be
$\{p_i(t): i \in S\}$, where

(1.1.3) $p_i(t) = \text{pr}\{X(t) = i\}$, $\sum_i p_i(t) = 1.$

The absolute distribution at time $t = 0$ is called the *initial distribution*. We have
the obvious relation

(1.1.4) $p_j(t) = \sum_i p_i(0) p_{ij}(t)$ $t \geq 0.$

The denumerable array of functions $(p_{ij}(.))$, $i, j \in S$, is the *transition matrix*
of the Markov process. It satisfies for every i, j and s, t the conditions

(1.1.5) $p_{ij}(t) \geq 0$

(1.1.6) $\sum_j p_{ij}(t) = 1$

(1.1.7) $\sum_k p_{ik}(s) p_{kj}(t) = p_{ij}(t+s).$

Conversely, any array of functions $(p_{ij}(.))$ satisfying (1.1.5) - (1.1.7) and
the initial condition $p_{ij}(0) = \delta_{ij}$ (δ_{ij} is Kroneckers's delta) for every $i, j \in S$
and s, $t \geq 0$, is the transition matrix of a Markov process $\{X(t): 0 \leq t < \infty\}$

for which (1.1.2) holds.

A transition matrix is called *standard* iff

(1.1.8) $\lim_{t \downarrow 0} p_{ij}(t) = p_{ij}(0) \equiv \delta_{ij}.$

A standard transition matrix has the property that each $p_{ij}(.)$ is uniformly continuous. Furthermore, the right-hand derivatives

(1.1.9) $q_{ij} = p_{ij}'(0) = \lim_{t \downarrow 0} (p_{ij}(t) - \delta_{ij})/t$

($p_{ij}'(t)$ will denote the right-hand derivative when $t = 0$, and the two-sided derivative when $t > 0$) exist, and they are finite except possibly when $i = j$. Always

(1.1.10) $0 \le q_{ij} < \infty$ $i \neq j,$

(1.1.11) $\sum_{j \neq i} q_{ij} \le -q_{ii} \le \infty;$

and in most cases of practical interest

(1.1.12) $q_{ii} > -\infty$ $i \in S,$

(1.1.13) $\sum_j q_{ij} = 0$ $i \in S.$

The matrix $(q_{ij}) = (p_{ij}'(0))$ is the *q-matrix* of the transition matrix $(p_{ij}(.))$; it is *stable* iff (1.1.12) holds and *conservative* iff it is stable and (1.1.13) holds. The elements $p_{ij}(.)$ of a stable transition matrix, i.e., a transition matrix having a stable q-matrix, have continuous derivatives. The functions $p_{ij}(.)$ satisfy the *backward equations*

(1.1.14) $p_{ij}'(t) = \sum_k q_{ik} p_{kj}(t)$

iff the q-matrix is conservative.

Given a conservative q-matrix, i.e., a matrix (q_{ij}) with non-negative elements off the main diagonal and satisfying both (1.1.12) and (1.1.13), then there exists at least one standard transition matrix $(p_{ij}(.))$ for which $p_{ij}'(0) = q_{ij}$, but in general this transition matrix is not unique. When it is unique the conservative q-matrix will be called *normal*.

For the elements $p_{ij}(.)$ of a stable transition matrix to satisfy the *forward equations*

(1.1.15) $p_{ij}'(t) = \sum_k p_{ik}(t) q_{kj}$

it is sufficient (but not necessary) that the q-matrix is normal.

Given a transition matrix $(p_{ij}(.))$, $i,j \in S$, the state i is called *absorbing* iff $p_{ii}(t) = 1$ for all $t > 0$. A necessary and sufficient condition for this is

(1.1.16) $q_{ii} = 0.$

The state i is *recurrent* iff

(1.1.17) $\int_0^\infty p_{ii}(t) = \infty,$

otherwise it is called *transient*. A recurrent state i is called *positive* or *null* according as $p_{ii} > 0$ or $p_{ii} = 0$, where

(1.1.18) $p_{ij} = \lim_{t \to \infty} p_{ij}(t).$

The latter limit exists for every i and j. A Markov process with transition matrix $(p_{ij}(.))$ is called *transient* (*null recurrent, positive recurrent*) iff every state $i \in S$ is transient (null recurrent, positive recurrent). The process is said to be *irreducible* on $S' \subset S$ when $p_{ij}(t) > 0$ for all $i,j \in S'$ and $t > 0$. It was shown by KENDALL (1959) that a Markov process with normal q-matrix has $p_{ij}(t) > 0$ for all $t > 0$ iff there exists a finite sequence $(i,k_1,...,k_r,j)$ with $r > 0$ and satisfying

(1.1.19) $q_{ik_1} q_{k_1 k_2} \cdots q_{k_r j} > 0.$

1.2   Stochastic monotonicity

We define R as the set of probability distribution vectors on $E \equiv \{0,1,2,...\}$, i.e.,

(1.2.1) $R \equiv \{\underline{r} = (r_0, r_1, ...)^T | \ r_i \geq 0 \ \& \ \Sigma \ r_i = 1\}$

(superscript T will denote transpose).

DEFINITION 1.2.1. Let $\underline{r}^{(1)}, \underline{r}^{(2)} \in R$. Then $\underline{r}^{(1)}$ *dominates* $\underline{r}^{(2)}$ $(\underline{r}^{(1)} \circ \geq \underline{r}^{(2)})$ iff for $i = 1,2,...$

(1.2.2) $\sum_{j \geq i} r_j^{(1)} \geq \sum_{j \geq i} r_j^{(2)} .$

The vector $\underline{r}^{(1)}$ *strictly dominates* $\underline{r}^{(2)}$ $(\underline{r}^{(1)} \circ > \underline{r}^{(2)})$ iff strict inequality holds in (1.2.2) for $i = 1,2,... $.

DEFINITION 1.2.2. An operator L mapping R into R is *monotone* iff for every pair $\underline{r}^{(1)}, \underline{r}^{(2)} \in R$ with $\underline{r}^{(1)} \circ > \underline{r}^{(2)}$

(1.2.3) $L(\underline{r}^{(1)}) \circ > L(\underline{r}^{(2)}).$

Now let {X(t): $0 \le t < \infty$} be a Markov process with state space
$S = E \equiv \{0, 1, \ldots \}$, say. The transition matrix $(p_{ij}(.))$ defines a set (in fact,
a semigroup) of operators P_t, $t \ge 0$, mapping R into R by means of

(1.2.4) $(P_t(\underline{r}))_j = \sum_i r_i p_{ij}(t)$, $\underline{r} = (r_0, r_1, \ldots)^T \in R$.

{P_t: $0 \le t < \infty$} is the set of *transition operators* of {X(t)}. If one denotes the
probability distribution vector of {X(t)} at time $t \ge 0$ by

(1.2.5) $\underline{p}(t) = (p_0(t), p_1(t), \ldots)^T$,

the relation (1.1.4) may be written as

(1.2.6) $\underline{p}(t) = P_t(\underline{p}(0))$.

More generally one has for s, $t \ge 0$

(1.2.7) $\underline{p}(t+s) = P_s(\underline{p}(t))$,

as can easily be verified.
In view of definition 1.2.1 it is natural to introduce the concept of stochastic
monotonicity in terms of {X(t)} as follows.

DEFINITION 1.2.3. The process {X(t)} is *stochastically increasing (decreasing)*
on the interval (t_1, t_2) iff for every pair τ_1, τ_2 with $0 \le t_1 \le \tau_1 < \tau_2 < t_2 \le \infty$

(1.2.8) $\underline{p}(\tau_2) \ {}^\circ{\ge}\ \underline{p}(\tau_1)$ $(\underline{p}(\tau_1) \ {}^\circ{\ge}\ \underline{p}(\tau_2))$.

The process is *strictly stochastically increasing (decreasing)* iff strict
domination prevails throughout.

It appears from the next theorem that in the context of stochastic monotonicity,
monotone transition operators are of particular interest.

THEOREM 1.2.4 (STOYAN (1977), Satz 4.2.4b). *The Markov process {X(t)} is
stochastically increasing (decreasing) on the interval (t_1, ∞) if the transition
operators P_t, $t \ge 0$, of {X(t)} are monotone and there exists a number $\tau > 0$
such that {X(t)} is stochastically increasing (decreasing) on the interval
$(t_1, t_1 + \tau)$.*

Several authors have given necessary and sufficient conditions for the transition
operator P_t to be monotone for all $t \ge 0$ (KEILSON and KESTER (1977), KIRSTEIN (1976),
STOYAN (1977)). The following is STOYAN's result.

THEOREM 1.2.5 (STOYAN (1977), Satz 4.2.9). *Let* {X(t)} *be a Markov process with state space* S = E *and normal q-matrix* (q_{ij}). *The transition operators* P_t, $t \geq 0$, *of* {X(t)} *are monotone iff*

(1.2.9) $$\sum_{j \geq k} q_{ij} \leq \sum_{j \geq k} q_{mj}$$

for all i,m *with* $i \leq m$ *and all* $k \leq i$ *and* $k > m$.

We remark that monotone operators occur widely in Markov processes (cf. KEILSON and KESTER (1977)), which is clearly an impetus for studying the phenomenon of stochastic monotonicity in the context of these processes.

This section is concluded with some notation. We define the infinite matrices $U = (u_{ij})$ and $V = (v_{ij})$, i, j = 0, 1, ..., by

(1.2.10) $u_{ij} = 1$ if $j \geq i$, 0 otherwise,

and

(1.2.11) $v_{ij} = 1$ if $j = i$, -1 if $j = i + 1$, 0 otherwise.

Then

(1.2.12) $UV = VU = I$,

the infinite identity matrix. Moreover, with $Q = (q_{ij})$, i, j = 0, 1, ..., it is readily seen that the necessary and sufficient condition expressed in theorem 1.2.5 can be formulated more simply as

(1.2.13) $(VQU)_{ij} \geq 0$ $\quad\quad\quad\quad i \neq j$,

or, alternatively, as

(1.2.14) $(V^T Q U^T)_{ij} \geq 0$ $\quad\quad\quad\quad i \neq j$.

For two vectors $\underline{a} = (a_0, a_1, \ldots)^T$ and $\underline{b} = (b_0, b_1, \ldots)^T$ we define

(1.2.15) $\underline{a} \leq \underline{b}$ iff $a_i \leq b_i$ for all i

and

(1.2.16) $\underline{a} < \underline{b}$ iff $a_i < b_i$ for all i.

In terms of the matrix U and the above forms of vector inequality, (strict) domination in R can be expressed as

(1.2.17) $\underline{r}^{(1)} \circ \geq \underline{r}^{(2)}$ iff $(\underline{r}^{(1)})^T U \leq (\underline{r}^{(2)})^T U$

and

(1.2.18) $\underline{r}^{(1)} \circ> \underline{r}^{(2)}$ iff $(\underline{r}^{(1)})^T U < (\underline{r}^{(2)})^T U,$

where we have used the fact that $\sum_i r_i^{(j)} = 1$. Thus a process is stochastically increasing (decreasing) on (t_1, t_2) iff for every pair τ_1, τ_2 with $0 \le t_1 \le \tau_1 < \tau_2 < t_2 \le \infty,$

(1.2.19) $\underline{p}^T(\tau_2)U \le \underline{p}^T(\tau_1)U \quad (\underline{p}^T(\tau_1)U \le \underline{p}^T(\tau_2)U)$

and strictly stochastically increasing (decreasing) iff strict inequality prevails, i.e.,

(1.2.20) $\underline{p}^T(\tau_2)U < \underline{p}^T(\tau_1)U \quad (\underline{p}^T(\tau_1)U < \underline{p}^T(\tau_2)U).$

1.3   Birth-death processes

It is customary to define a *birth-death process* as a Markov process with state space $S = \{-1, 0, 1, \ldots\}$, say (including a state labeled -1 will appear to have notational advantages), and transition probability functions $p_{ij}(.)$ which satisfy the conditions

(1.3.1)
$$\begin{aligned} p_{i,i+1}(t) &= \lambda_i t + o(t) \\ p_{ii}(t) &= 1 - (\lambda_i + \mu_i)t + o(t) \\ p_{i,i-1}(t) &= \mu_i t + o(t) \end{aligned}$$

as $t \to 0$, where $\lambda_{-1} = \mu_{-1} = 0$, $0 \le \mu_0 < \infty$, $0 < \mu_i < \infty$ for $i > 0$ and $0 < \lambda_i < \infty$ for $i \ge 0$ (see, e.g., FELLER (1967)). A number of authors (DOBRUŠIN (1952), FELLER (1959), KENDALL and REUTER (1957), REUTER (1957)) have shown that the postulates (1.3.1) uniquely determine the transition matrix of the process iff the parameters λ_i and μ_i satisfy the condition

(1.3.2) $\sum_{n=0}^{\infty}(1/\lambda_n \pi_n) \sum_{i=0}^{n} \pi_i = \infty,$

where the *potential coefficients* π_n are given by

(1.3.3) $\pi_0 = 1; \quad \pi_n = \lambda_0 \ldots \lambda_{n-1} / \mu_1 \ldots \mu_n$ $n > 0.$

As a direct consequence of (1.3.1) one has

(1.3.4) $\lim_{t \to 0} p_{ij}(t) = \delta_{ij},$

i.e., the transition matrix of a birth-death process is standard. Furthermore, it is not difficult to show that a Markov process satisfying (1.3.1), will also satisfy the differential equations

$$(1.3.5) \qquad p'_{ij}(t) = \sum_k a_{ik} p_{kj}(t) \qquad\qquad i,j \in S, \ t \geq 0,$$

where

$$(1.3.6) \qquad a_{ij} = \lambda_i \qquad \text{if } j = i+1$$
$$= -(\lambda_i + \mu_i) \quad \text{if } j = i$$
$$= \mu_i \qquad \text{if } j = i-1$$
$$= 0 \qquad \text{otherwise} \cdot$$

In particular

$$(1.3.7) \qquad p'_{ij}(0) = a_{ij} \cdot$$

Hence the matrix (a_{ij}) defined by (1.3.6) is the (conservative) q-matrix of the birth-death process with parameters λ_i and μ_i, and (1.3.5) are the backward equations. Conversely, it is easy to see that a Markov process with a conservative q-matrix of the type (1.3.6) will satisfy the relations (1.3.1).

So postulating (1.3.1) amounts to prescribing the q-matrix (1.3.6). Apparently, the conservative q-matrix (a_{ij}) is normal iff (1.3.2) is satisfied. KEMPERMAN (1962) has shown that of the many properties of the transition matrix going with a given normal q-matrix $(q_{ij}) = (a_{ij})$, to wit (1.1.5)-(1.1.8), (1.1.14), (1.1.15) and the initial condition

$$(1.3.8) \qquad p_{ij}(0) = \delta_{ij},$$

merely two uniquely determine this transition matrix. Namely, (1.3.2) is necessary and sufficient for the equations (1.3.5) (=(1.1.14)) with initial condition (1.3.8) to have a unique solution. As we have seen in section 1.1 this unique solution will satisfy the forward equations

$$(1.3.9) \qquad p'_{ij}(t) = \sum_k p_{ik}(t) a_{kj} \cdot$$

KEMPERMAN (1962) has proved a converse statement too. Namely, the condition

$$(1.3.10) \qquad \sum_{n=0}^{\infty} (1/\lambda_n \pi_n) \sum_{i=n+1}^{\infty} \pi_i = \infty$$

is necessary and sufficient for the equations (1.3.9) with initial condition (1.3.8) to have a unique solution, and this solution will satisfy the backward equations (1.3.5).

In their studies of birth-death processes KARLIN and McGREGOR (1957[a], 1957[b]) take
as a starting point the differential equations (1.3.5) and (1.3.9) with initial
condition (1.3.8). KEMPERMAN's results imply that this system of equations has
exactly one solution if (1.3.2) or (1.3.10) is satisfied; equivalently,

$$(1.3.11) \qquad \sum_{n=0}^{\infty} (\pi_n + 1/\lambda_n \pi_n) = \infty.$$

KARLIN and McGREGOR (1957[a]) have shown that this condition is also necessary.

It is easily verified that the q-matrix of a birth-death process satisfies
condition (1.2.9) of theorem 1.2.5. Thus, provided (1.3.2) is satisfied (the
q-matrix is normal), the transition operators of a birth-death process are monotone.
As a consequence of this fact and theorem 1.2.4 the following holds true.

THEOREM 1.3.1. *A birth-death process with parameters λ_i and μ_i satisfying
(1.3.2) is stochastically increasing (decreasing) on the interval (t_1, ∞) iff
there exists a $\tau > 0$ such that it is stochastically increasing (decreasing) on the
interval $(t_1, t_1+\tau)$.*

1.4   Some notation and terminology

Any set of numbers $\{\lambda_n, \mu_n : n = 0, 1, \ldots \ldots\}$ satisfying

$$(1.4.1) \qquad \begin{array}{ll} 0 \le \mu_0 < \infty & \\ 0 < \lambda_n < \infty & n \ge 0 \\ 0 < \mu_n < \infty & n > 0 \end{array}$$

is called *a set of birth-death parameters.* Associated with a set of birth-death
parameters $\{\lambda_n, \mu_n\}$ is the matrix $A = (a_{ij})$, i, j = 0, 1,...., defined by

$$(1.4.2) \qquad A = \begin{pmatrix} -(\lambda_0+\mu_0) & \lambda_0 & 0 & \cdots \\ \mu_1 & -(\lambda_1+\mu_1) & \lambda_1 & \cdots \\ 0 & \mu_2 & -(\lambda_2+\mu_2) & \cdots \\ \cdots & \cdots & \cdots & \cdots \\ \cdots & \cdots & \cdots & \cdots \end{pmatrix} .$$

A is called a *generator.* It is seen that A is in fact the q-matrix of the
birth-death process with parameters $\{\lambda_n, \mu_n\}$ as defined in the previous section,
after deletion of the first row and column (corresponding to the state -1).

In terms of a generator A the conditions (1.3.2) and (1.3.10) are denoted by C(A) and D(A), respectively, i.e.,

(1.4.3) $C(A)$ iff $\sum_{n=0}^{\infty} (1/\lambda_n \pi_n) \sum_{i=0}^{n} \pi_i = \infty$

and

(1.4.4) $D(A)$ iff $\sum_{n=0}^{\infty} (1/\lambda_n \pi_n) \sum_{i=n+1}^{\infty} \pi_i = \infty.$

It will be convenient to describe birth-death processes which are determined by a generator A satisfying C(A) and D(A), as *natural*, with reference to their "boundaries at infinity" (see FELLER (1959), CALLAERT and KEILSON (1973)). The transition probability functions $p_{ij}(.)$ of a natural birth-death process satisfy the backward equations, whence in particular

$$p'_{-1,j}(t) = 0,$$

so that

(1.4.5) $p_{-1,j}(t) = \delta_{-1,j}.$

The forward equations imply

$$p'_{i,-1}(t) = \mu_0 p_{i0}(t),$$

whence

(1.4.6) $p_{i,-1}(t) = \mu_0 \int_0^t p_{i0}(\tau) d\tau$ $i \geq 0.$

Considering (1.4.5) and (1.4.6) we shall not be bothered much about the state -1 in what follows. Actually, when $\mu_0 = 0$ this state is completely irrelevant (0 is a reflecting barrier) and we shall mean the set $E \equiv \{0, 1, ...\}$ when we talk about the state space of a birth-death process where $\mu_0 = 0$. For $t \geq 0$ we define the matrix $P(t) = (p_{ij}(t))$, $i, j = 0, 1, ... $. In terms of $P(.)$ and the associated generator A the backward equations (1.3.5) reduce to

(1.4.7) $P'(t) = AP(t),$

while the forward equations (1.3.9) become

(1.4.8) $P'(t) = P(t)A.$

Obviously, one has the initial condition

(1.4.9) $P(0) = I.$

If we let

(1.4.10) $\underline{p}(t) = (p_0(t), p_1(t), \ldots)^T,$

the relation (1.1.4) for $j = 0, 1, \ldots$ may be represented as

(1.4.11) $\underline{p}(t) = P^T(t)\underline{p}(0).$

More generally one has

(1.4.12) $\underline{p}(t+s) = P^T(s)\underline{p}(t).$

There is an obvious correspondence between (1.2.6), (1.2 7) on the one hand and (1.4.11), (1.4.12) on the other: we have dropped the operator notation in favour of the more tractable matrix notation.
For completeness we note that, since $\sum\limits_{j=-1}^{\infty} p_j(t) = 1$, one has

(1.4.13) $p_{-1}(t) = 1 - \underline{p}^T(t)\underline{1},$

$\underline{1}$ denoting the infinite column vector consisting of 1's. One readily verifies

(1.4.14) $\underline{p}^T(t)\underline{1} = \underline{p}^T(0)\underline{1}$ if $\mu_0 = 0.$

To avoid inessential difficulties it will be assumed throughout that $\underline{p}^T(0)\underline{1} = 1$ if $\mu_0 = 0$.
Finally, we remark that the condition (1.1.19) is satisfied for $i, j = 0, 1, \ldots$, when dealing with a birth-death process with parameters (1.4.1). So a natural birth-death process is irreducible on E.

2.   NATURAL BIRTH-DEATH PROCESSES

2.1   Some basic properties

Let $\{\lambda_n, \mu_n\}$ be a set of birth-death parameters, A the associated generator (1.4.2)
and $\{X(t): 0 \le t < \infty\}$ a birth-death process with generator A. In this and the
following chapters we shall be concerned with natural birth-death processes only,
i.e., A is assumed to satisfy the conditions C(A) and D(A). The state -1 will
be disregarded and the term transition matrix will be used for the matrix
$P(.) = (p_{ij}(.))$, where i, j $= 0, 1, \ldots\ldots$. Since the properties to be discussed
in this chapter are independent of the initial distribution of the process, we shall
often identify the birth-death process $\{X(t)\}$ with its transition matrix $P(.)$.
KARLIN and McGREGOR (1957[a], 1959[b]) have proved the following important feature
of the transition matrix $P(.)$ of a natural birth-death process:

(2.1.1) $P(t)$ is *strictly totally positive* (*STP*) for $t > 0$,

which means that every subdeterminant of $P(t)$ is strictly positive for $t > 0$.
KARLIN and McGREGOR (1959[a]) showed that property (2.1.1) is characteristic for
birth-death processes in the class of Markov processes as defined in section
1.1. An immediate conclusion from (2.1.1) is

(2.1.2) $p_{ij}(t) > 0$ i, j $= 0, 1, \ldots$; $t > 0$.

Before giving another consequence of (2.1.1), which in fact will appear to be
an important tool when obtaining the results of chapter 5, we need some
preliminaries.
A sequence $(x_0, x_1, x_2, \ldots)$ is said to have a *change of sign at* k if j,
the first index $> k$ for which $x_j \neq 0$, exists and $x_k x_j < 0$. For any vector
$\underline{x} = (x_0, x_1, \ldots)^T \neq \underline{0}$ (the latter denoting the column vector consisting of 0's),
$S^-(\underline{x})$ denotes the number of sign changes in the sequence $(x_0, x_1, \ldots)$. Furthermore,
$S^+(\underline{x})$ is the maximum number of sign changes possible in the sequence
$(x_0, x_1, \ldots)$ by allowing each zero to be replaced by ± 1. The two theorems on
STP-matrices in appendix 1 and (2.1.1) imply the next theorem.

THEOREM 2.1.1. *The transition matrix* $P(t) = (p_{ij}(t))$, $i, j = 0, 1, \ldots,$ *of a natural birth-death process has for* $t > 0$ *the property*

(2.1.3) $S^{+}(P(t)\underline{x}) \leq S^{-}(\underline{x})$

for any vector $\underline{x} = (x_0, x_1, \ldots)^T \neq \underline{0}$ *such that* $P(t)\underline{x}$ *exists. Moreover, if* $S^{-}(P(t)\underline{x}) = S^{+}(P(t)\underline{x}) = S^{-}(\underline{x}) < \infty$, *then the first component of* $P(t)\underline{x}$ *is non-zero and its sign equals the sign of the first non-zero component of* $\underline{x}$.

REMARK 2.1.2. If the matrix P is STP then, obviously, also P^T is STP. Consequently, the sign variation diminishing property of the transition matrix $P(t)$ holds under post-multiplication but also under pre-multiplication by a vector $\underline{x}$ such that $P(t)\underline{x}$, respectively $P^T(t)\underline{x} = (\underline{x}^T P(t))^T$, exists.

From (1.1.6) it is seen that for fixed i and t

(2.1.4) $p_{ij}(t) \to 0$ as $j \to \infty$.

It can also be shown that for fixed j and t

(2.1.5) $p_{ij}(t) \to 0$ as $i \to \infty$.

The proof of (2.1.5) will be given in section 3.2, since it is an easy consequence of our results on dual birth-death processes.
We conclude this section with the observation that the transition probabilities $p_{ij}(t)$, $i, j = 0, 1, \ldots,$ $t \geq 0$, of a natural birth-death process satisfy

(2.1.6) $\pi_i p_{ij}(t) = \pi_j p_{ji}(t)$

(KARLIN and McGREGOR (1957[b]), KENDALL (1959), REICH (1957)). It was shown by KENDALL (1959) that this *reversibility* property of birth-death processes makes that the transition probabilities have integral representations as given in the next section.

2.2   The spectral representation

Associated with a set $\{\lambda_n, \mu_n : n = 0, 1, \ldots\}$ of birth-death parameters (determining a generator A), is a system $\{Q_n(x) : n = 0, 1, \ldots\}$ of polynomials defined by the recurrence relations

$$Q_0(x) = 1$$
(2.2.1) $$-xQ_0(x) = -(\lambda_0 + \mu_0)Q_0(x) + \lambda_0 Q_1(x)$$
$$-xQ_n(x) = \mu_n Q_{n-1}(x) - (\lambda_n + \mu_n)Q_n(x) + \lambda_n Q_{n+1}(x) \qquad n \geq 1.$$

It has been shown by KARLIN and McGREGOR (1957[a]) that under the assumptions
C(A) and D(A) there is a unique, non-decreasing function $\psi(x)$ which is continuous
to the left and has $\psi(x) = 0$ for $x \leq 0$ and $\psi(x) \to 1$ as $x \to \infty$, such that

$$(2.2.2) \qquad \pi_j \int_0^\infty Q_i(x) Q_j(x) d\psi(x) = \delta_{ij}.$$

The function ψ is called the *spectral function* of the process which is determined
by A. The crux of KARLIN and McGREGOR's (1957[a]) results is that the transition
matrix $P(t) = (p_{ij}(t))$ is represented by the formula

$$(2.2.3) \qquad p_{ij}(t) = \pi_j \int_0^\infty \exp(-xt) Q_i(x) Q_j(x) d\psi(x).$$

The expression (2.2.2) exhibits the fact that the polynomials $Q_n(x)$,
$n = 0, 1, \ldots$, are orthogonal polynomials with respect to the spectral function ψ.
Consequently, the zeros of $Q_n(x)$, $n = 1, 2, \ldots$, are real and positive (SZEGÖ
(1959), theorem 3.3.1). Moreover, with $x_{1,n} \leq x_{2,n} \leq \ldots \leq x_{n,n}$ denoting the
zeros of $Q_n(x)$, one has

$$(2.2.4) \qquad x_{i-1,n} < x_{i,n+1} < x_{i,n} \qquad\qquad i = 1, 2, \ldots, n+1,$$

where $x_{0,n} \equiv 0$ and $x_{n+1,n} \equiv \infty$ (SZEGÖ (1959), theorem 3.3.2). Hence for fixed i,
$x_{i,n}$ tends to a limit x_i as n approaches infinity.
We define $S(\psi)$, the *spectrum* of ψ, as the set of all points x, such that for all
$\varepsilon > 0$, $\psi(x+\varepsilon) - \psi(x-\varepsilon) > 0$. According to KARLIN and McGREGOR (1957[a]), $S(\psi)$
consists of a countable infinity of points. There is an intimate relationship
between $S(\psi)$ and the zeros $\{x_{i,n}\}$ of the polynomials $\{Q_n(x)\}$, as appears from the
next lemmas.

LEMMA 2.2.1 (STONE(1964), theorem 10.42; SHOHAT and TAMARKIN (1963), lemma 4.3).
*Let L be the closed set on the real axis consisting of the accumulation points of
the set* $\{x_{i,n}: i = 1, 2, \ldots, n; n = 1, 2, \ldots\}$, *then* $S(\psi) \subset L$.

LEMMA 2.2.2 (SZEGÖ (1959), theorem 3.3.1). *If* $S(\psi) \subset [a,b]$, *then*
$\{x_{i,n}: i = 1, 2, \ldots, n; n = 1, 2, \ldots\} \subset (a,b)$.

LEMMA 2.2.3 (SZEGÖ (1959), theorem 3.41.2). *In the open interval between two
consecutive zeros of* $Q_n(x)$, *the spectral function* $\psi(x)$ *cannot be constant,
i.e.*, $S(\psi) \cap (x_{i,n}, x_{i+1,n}) \neq \phi$.

With the preceding results it is fairly easy to prove the next lemma,
which is due to CALLAERT.

LEMMA 2.2.4 (CALLAERT (1971), theorem (1.19)).
(i)   If $0 \leq x_1 < x_2$, then $x_1 \in S(\psi)$.
(ii)   If $k > 1$ and $x_{k-1} < x_k < x_{k+1}$, then $x_k \in S(\psi)$.
(iii)   If $k \geq 1$ and $x_k = x_{k+1}$, then $x_k \in S(\psi)$ and $S(\psi) \cap (x_k, x_k+\epsilon) \neq \emptyset$ for
        any $\epsilon > 0$.

Additionally the following holds.

LEMMA 2.2.5. If $k \geq 1$ and $x_{k+1} = x_k$, then $x_{k+n} = x_k$ for all natural n.

PROOF. Let $x_{k+1} = x_k$ and suppose $x_{k+2} > x_{k+1}$. Then, for n sufficiently large,

$$x_k = x_{k+1} < x_{k,n} < x_{k+1,n} < x_{k+2}$$

according to (2.2.4). Thus $Q_n(x)$ has two zeros in the interval (x_{k+1}, x_{k+2}),
whence $S(\psi) \cap (x_{k+1}, x_{k+2}) \neq \emptyset$ by lemma 2.2.3. This, however, contradicts lemma
2.2.1, since, obviously, $L \cap (x_{k+1}, x_{k+2}) = \emptyset$. The lemma follows by induction.   □

The above two lemmas and part of lemma 2.2.1 can be summarized in the next
theorem.

THEOREM 2.2.6. If the limit points x_i, $i = 1, 2, \ldots$, are all distinct, then
$S(\psi) \cap [0,x_\infty) = \{x_i : i = 1, 2, \ldots\}$, where $x_\infty = \lim_{i \to \infty} x_i$ (possibly infinity). If
the limit points are not all distinct, then there exists a positive integer
k such that $0 \leq x_1 < x_2 < \ldots < x_k$ and $x_{k+n} = x_k$ for all natural n; furthermore
$S(\psi) \cap [0,x_k] = \{x_i : i = 1, 2, \ldots\}$ and x_k is an accumulation point of elements
of $S(\psi)$.

    We define

(2.2.5) $\Delta\psi(x) \equiv \psi(x+0) - \psi(x-0) = \lim_{\epsilon \downarrow 0} \psi(x+\epsilon) - \psi(x)$.

The next theorem shows that $\psi(x)$ has a jump in an isolated limit point.

THEOREM 2.2.7. *If* $k \geq 1$ *and* $x_k < x_{k+1}$, *then* $\Delta\psi(x_k) > 0$.

PROOF. Let $0 < \varepsilon < x_{k+1} - x_k$ and, if $k > 1$, $\varepsilon < x_k - x_{k-1}$. Then, in view of the preceding theorem, ψ is constant on $(x_k, x_k + \varepsilon)$, so that $\Delta\psi(x_k) = \psi(x_k + \varepsilon) - \psi(x_k)$. Also, ψ is constant on $[x_k - \varepsilon, x_k)$, so that $\psi(x_k) - \psi(x_k - \varepsilon) = 0$. It follows that $\Delta\psi(x_k) = \psi(x_k + \varepsilon) - \psi(x_k - \varepsilon)$, which is strictly positive since $x_k \in S(\psi)$. □

Although the foregoing results give much insight into the behaviour of the spectral function ψ, they are of little use when one is interested in an exact expression for ψ. However, many models of birth-death processes that have been discussed in the literature are associated with a system of polynomials $\{Q_n(x)\}$, which, after some simple change of variable and renormalization, are classical polynomials (cf. chapter 8). For these processes the computation of the spectral function presents no problem. There are also important processes for which the corresponding polynomials do not reduce to classical polynomials but which can be tackled using a method described by KARLIN and McGREGOR (1957[b]). They suggest to solve the problem of calculating ψ by trying to find an explicit expression for the Stieltjes transform of ψ, viz.,

(2.2.6) $B(z) = \int_0^\infty d\psi(x)/(x-z)$ $0 < \arg z < 2\pi, \ |z| > 0$,

which is a holomorphic function in its domain of definition (WIDDER (1946)), and then computing the spectral function by means of the formula for inverting a Stieltjes transform, viz.,

(2.2.7) $\frac{1}{2}(\psi(x+0) + \psi(x)) = \lim_{\eta \downarrow 0} (1/\pi) \int_{-\varepsilon}^{x} \mathrm{Im}\, B(\xi + i\eta)\, d\xi$,

where $\varepsilon > 0$ (WIDDER (1946)). The problem of finding $B(z)$ can be solved by various techniques, examples of which will be referred to in the chapters 6 and 7. Once $B(z)$ is found it is usually a matter of routine to continue $B(z)$ into the complex plane; we shall denote the continued function by $H(z)$. The procedure of applying the inversion formula (2.2.7) has certain steps which are common to all spectral functions and which will be dealt with subsequently. Evidently, (2.2.7) may be written as

(2.2.8) $\frac{1}{2}(\psi(x+0) + \psi(x)) = \lim_{\eta \downarrow 0} (1/\pi) \int_{-\varepsilon}^{x} \mathrm{Im}\, H(\xi + i\eta)\, d\xi$.

It follows that

(2.2.9) $\frac{1}{2}(\psi(x+\delta+0) - \psi(x+0)) + \frac{1}{2}(\psi(x+\delta) - \psi(x)) =$

$= \frac{1}{2}(\psi(x+\delta+0) + \psi(x+\delta)) - \frac{1}{2}(\psi(x+0) + \psi(x)) =$

$= \lim_{\eta \downarrow 0} (1/\pi) \int_{x}^{x+\delta} \mathrm{Im}\, H(\xi + i\eta)\, d\xi$.

We note that the function B(z), and hence the function H(z), has conjugate values
at conjugate points (WIDDER(1946)), so that Im H(x) = 0 for any real number x
which is not a singularity of H(z). Furthermore, on any closed interval containing
no singularities of H(z), Im H(ξ+iη) tends uniformly to Im H(ξ) = 0 as η ↓ 0.
Thus if x is not a singularity of H(z) and $|δ|$ is sufficiently small then

$$\tfrac{1}{2}(\psi(x+\delta+0) - \psi(x+0))+ \tfrac{1}{2}(\psi(x+\delta) - \psi(x)) = 0,$$

whence in particular ψ(x+δ) = ψ(x), considering that ψ is non-decreasing.
Consequently, ψ(x) is constant, i.e., ψ'(x) = 0, between the singularities of H(z).
We next consider the case that x is a pole of H(z). Since H(z) has conjugate values
at conjugate points, we have

(2.2.10) $2i\,\mathrm{Im}\,H(\xi+i\eta) = H(\xi+i\eta) - H(\xi-i\eta),$

so that

$$(1/\pi) \int_{-\varepsilon}^{x} \mathrm{Im}\,H(\xi+i\eta)\,d\xi = (1/2\pi i) \left\{ \int_{-\varepsilon+i\eta}^{x+i\eta} H(\xi)\,d\xi + \int_{-x-i\eta}^{-\varepsilon-i\eta} H(\xi)\,d\xi \right\}.$$

Hence for δ > 0 sufficiently small

(2.2.11) $$\psi(x+\delta) - \psi(x-\delta) = \lim_{\eta\downarrow 0}(1/2\pi i) \left\{ \int_{x-\delta+i\eta}^{x+\delta+i\eta} H(\xi)\,d\xi + \int_{x+\delta-i\eta}^{x-\delta-i\eta} H(\xi)\,d\xi \right\}.$$

One readily verifies by contour integration that the right member of (2.2.11)
reduces to $-\mathrm{Res}_x H(z)$: the residu of $-H(z)$ at the pole x. Therefore, ψ has a
jump of magnitude $\Delta\psi(x) = -\mathrm{Res}_x H(z)$ at the point x if x is a pole of H(z).
Summarizing our results we have the following theorem.

THEOREM 2.2.8. *With H(z) denoting the analytic continuation of the Stieltjes
transform of a spectral function ψ, one has ψ'(x) = 0 if x is not a
singularity of H(z) and Δψ(x) = -Res$_x$H(z) if x is a pole of H(z).*

It is interesting to compare the theorems 2.2.7 and 2.2.8. It then appears that
an isolated limit point x_i is a singularity of the function H(z).

2.3 Exponential ergodicity

Let $\{X(t) : 0 \leq t < \infty\}$ be a temporally homogeneous Markov process on the denumerable state space S. The probability $p_{ij}(t)$ of (1.1.2) is said to go *exponentially fast* to its limits p_{ij} as $t \to \infty$ iff

$$p_{ij}(t) - p_{ij} = O(\exp(-\alpha_{ij}t))$$

for some $\alpha_{ij} > 0$. If $p_{ij}(t)$ goes to its limit exponentially fast, then the *decay parameter* $\hat{\alpha}_{ij}$ of $p_{ij}(t)$ is defined by

(2.3.1) $\hat{\alpha}_{ij} = \sup\{\alpha_{ij}|p_{ij}(t) - p_{ij} = O(\exp(-\alpha_{ij}(t))\}.$

KINGMAN (1963[a], 1963[b]) has shown that exponential decay is a solidarity property, i.e., if the process is irreducible on $S' \subset S$, then either all or none of the transition probabilities $p_{ij}(t)$, i, $j \in S'$, go exponentially fast to their limits. If they do the process is called *exponentially ergodic on* S' and there exists a positive *common decay parameter*

(2.3.2) $\hat{\alpha} = \sup\{\alpha|p_{ij}(t) - p_{ij} = O(\exp(-\alpha t))$ for all i, $j \in S'\}.$

CALLAERT (1971, 1974) has studied the phenomenon of exponential ergodicity in the context of birth-death processes. His main conclusions are given in the next theorem. As usual only natural birth-death processes are considered, which are irreducible on $E \equiv \{0, 1, \ldots\}$.

THEOREM 2.3.1 (CALLAERT (1974), theorems 1 and 2). *Let* x_i, i = 1, 2, ..., *be the limit points associated with a given birth-death process with* $\mu_0 = 0$. *Then the decay parameter* $\hat{\alpha}_{00}$ *is given by*

$$\hat{\alpha}_{00} = x_1 \ if \ x_1 > 0, \ x_2 \ if \ x_2 > x_1 = 0;$$

$P_{00}(t)$ *does not go to its limit exponentially fast if* $x_1 = x_2 = 0$.
If the process is exponentially ergodic, i.e., $x_2 > 0$, *then the common decay parameter* $\hat{\alpha}$ *is equal to* $\hat{\alpha}_{00}$. *If* $\hat{\alpha}_{00} = x_1 > 0$, *then* $\hat{\alpha}_{ij} = \hat{\alpha}$ *for all* i, j = 0, 1, ... *If* $x_1 = 0$ *and* $x_2 > 0$, *then*

$$\hat{\alpha}_{ij} = \hat{\alpha}_{ji} \geq (\hat{\alpha}_{ii} + \hat{\alpha}_{jj})/2 \geq \hat{\alpha}$$

for all i, j. *Moreover, if* $\hat{\alpha}_{ii} > \hat{\alpha}$, *then* $\hat{\alpha}_{nn} = \hat{\alpha}$ *for all* $n \neq i$.

To investigate exponential ergodicity and the decay parameter of a birth-death process with $\mu_0 > 0$, CALLAERT used the transformation discussed in the next chapter. The following theorem is an easy consequence of CALLAERT's (1974) theorem 3 and the results (3.1.3), (3.1.4), (3.3.12) – (3.3.14).

THEOREM 2.3.2. *Let* x_i, *i* = 1,2,..., *be the limit points associated with a given birth-death process with* $\mu_0 > 0$. *Then the common decay parameter* â *is given by*

$$\hat{a} = x_1 \ if \ x_1 > 0;$$

the probabilities $p_{ij}(t)$, *i, j* = 0, 1,....., *do not go to their limits exponentially fast if* $x_1 = 0$. *If the process is exponentially ergodic then* $\hat{a}_{ij} = \hat{a} = x_1$ *for all i, j* = 0, 1, ...

Anticipating (3.3.15) and (3.3.16) we remark that $x_2 = 0$ if $\mu_0 > 0$ and $x_1 = 0$. Considering this fact and theorem 2.2.6, it follows from the preceding two theorems that a natural birth-death process is exponentially ergodic iff the first point > 0 of its spectrum exists. Moreover, this point, if existing, is the common decay parameter. Evidently, nonexistence of the point > 0 in the spectrum of a natural birth-death process is equivalent to 0 being a point of accumulation of the spectrum.

2.4   The moment problem and related topics

The following problem is known as the *Stieltjes moment problem:*
Find a bounded non-decreasing function $\psi(x)$ on the interval $[0,\infty)$ such that its *moments* $\int_0^\infty x^n d\psi(x)$, n = 0, 1, ..., have a prescribed set of values

$$(2.4.1) \qquad \int_0^\infty x^n d\psi(x) = m_n \qquad\qquad n = 0, 1, ...$$

(For convenience we shall consider functions ψ which are continuous to the left and have $\psi(0) = 0$).
It was KARLIN and McGREGORS's brilliant idea to transform the problem of finding the spectral function of a birth-death process into a Stieltjes moment problem. This was done as follows.
Let $\{\lambda_n, \mu_n\}$ be the set of birth-death parameters of a natural birth-death process and $\{Q_n\}$ the associated set of polynomials (2.2.1). The system of equations $\int Q_0 d\psi = 1$, $\int Q_n d\psi = 0$, n > 0, can be solved recursively for the moments $c_n = \int x^n d\psi$. For example $\int Q_0 d\psi = 1$ gives $c_0 = 1$, and then

$$0 = \int Q_1 d\psi = (1/\lambda_0) \int (\lambda_0 + \mu_0 - x) d\psi$$

gives $c_1 = \lambda_0 + \mu_0$. Obviously, the sequence c_n, $n = 0, 1, \ldots$, is uniquely associated with the set $\{\lambda_n, \mu_n\}$ and consequently there is a unique Stieljes moment problem associated with the set $\{\lambda_n, \mu_n\}$, viz., $m_n = c_n$, $n = 0, 1, \ldots$, in (2.4.1). The fact that we consider natural processes only is sufficient for the associated moment problem to have a unique solution; in addition this unique solution is precisely the spectral function of the birth-death process (KARLIN and McGREGOR (1957[a])).

The connection between birth-death processes and the Stieltjes moment problem makes that many results from the theory of the latter problem can be transformed into a result about birth-death processes. In this way KARLIN and McGREGOR solved the problems of existence and uniqueness of birth-death processes, given a set of parameters. As a second example we mention lemma 2.2.1, which is a consequence of lemma 4.3 in SHOHAT and TAMARKIN's (1963) book about the moment problem. A third example will be given shortly. It concerns the question under which conditions the spectrum of a natural birth-death process will be concentrated on a finite segment. CALLAERT and KEILSON (1973), section 11, took some pains to solve this problem but they failed to find the simple necessary and sufficient condition. We settle the question by employing a theorem of TOEPLITZ, given in AHIEZER and KREIN (1962), article vi, theorem 1. Using the findings of KARLIN and McGREGOR (1957[a]) on the moment sequence associated with a natural birth-death process this theorem implies the following.

LEMMA 2.4.1. *Let $\{\lambda_n, \mu_n\}$ be a set of birth-death parameters of a natural birth-death process and $\{c_n\}$ the associated moments. In order that the unique solution ψ of the Stieltjes moment problem corresponding to $\{c_n\}$ is such that all growth points (i.e., the spectrum) are concentrated in a finite segment, it is necessary and sufficient that the set of numbers.*

(2.4.2) $a_k = (D_k'/D_k) - (D_{k-1}'/D_{k-1})$, $b_k = \sqrt{D_{k-1}D_{k+1}}/D_k$ $k = 0, 1, \ldots$

is bounded, where

(2.4.3) $D_{-1} = 1$, $D_k = \begin{vmatrix} c_0 & \cdots & c_k \\ \vdots & & \vdots \\ c_k & \cdots & c_k \end{vmatrix}$ $k = 0, 1, \ldots$

and

$$(2.4.4) \qquad D'_{-1} = 0, \ D'_0 = c_1, D'_k = \begin{vmatrix} c_0 & c_1 & \cdot & \cdot & c_{k-1} & c_{k+1} \\ c_1 & c_2 & \cdot & \cdot & c_k & c_{k+2} \\ \cdot & \cdot & \cdot & \cdot & \cdot & \cdot & \cdot & \cdot & \cdot \\ c_k & c_{k+1} & \cdot & \cdot & c_{2k-1} & c_{2k+1} \end{vmatrix} \qquad k > 0.$$

According to our previous statement (2.2.2) the polynomials Q_n, $n = 0, 1, \ldots,$ associated with a natural birth-death process are orthogonal with respect to the spectral function ψ. If we normalize these polynomials such that the resulting polynomials P_n have a positive leading coefficient and

$$(2.4.5) \qquad \int P_i(x) P_j(x) \, d\psi(x) = \delta_{ij},$$

it is readily seen that

$$(2.4.6) \qquad P_n(x) = (-1)^n \sqrt{\pi_n} Q_n(x).$$

Comparison of the recursive relations for $P_n(x)$ resulting from (2.4.6) and (2.2.1) with those found by AKHIEZER (1965), section 1.3, for the normalized orthogonal polynomials belonging to ψ reveals that

$$(2.4.7) \qquad a_k = \lambda_k + \mu_k, \ b_k = \sqrt{\lambda_k \mu_{k+1}} \qquad k = 0, 1, \ldots.$$

Lemma (2.4.1) and (2.4.7) lead to the next theorem.

THEOREM 2.4.2. *The spectrum of a natural birth-death process with parameters* $\{\lambda_n, \ \mu_n\}$ *is concentrated on a finite segment if and only if* $\sup\{\lambda_k + \mu_k\} < \infty$.

It is well known that a close connection exists between the moment problem and spectral theory of operators (see AKHIEZER (1965), chapter 4). For a given set of moments $\{c_n\}$ the link is established by the Jacobi matrix

$$(2.4.8) \qquad \begin{pmatrix} a_0 & b_0 & 0 & 0 & 0 & \cdot & \cdot & \cdot & \cdot & \cdot \\ b_0 & a_1 & b_1 & 0 & 0 & \cdot & \cdot & \cdot & \cdot & \cdot \\ 0 & b_1 & a_2 & b_2 & 0 & \cdot & \cdot & \cdot & \cdot & \cdot \\ \cdot & \cdot & \cdot & \cdot & \cdot & \cdot & \cdot & \cdot & \cdot & \cdot & \cdot \\ \cdot & \cdot & \cdot & \cdot & \cdot & \cdot & \cdot & \cdot & \cdot & \cdot & \cdot \end{pmatrix} ,$$

where a_k and b_k are given by (2.4.2). This matrix can be considered as the matrix of a linear operator in Hilbert space. For details we refer to

AKHIEZER (1965), BEREZANSKIĬ (1968) and STONE (1964).

Considering our previous remarks regarding the relation between a set
of birth-death parameters and a Stieltjes moment problem, there will also
exist a relation between the birth-death process with parameters $\{\lambda_n, \mu_n\}$ and
the operator defined by the corresponding (via the moments $\{c_n\}$) Jacobi matrix
(2.4.8). This explains the fact that lemma 2.2.1 can also be found in STONE's
(1964) book about linear operators.

In conclusion of this section we mention the interesting works of CASE (1974, 1975)
and MAKI (1976). CASE states amongst other things that if the numbers a_k and b_k
defined by (2.4.7) have limits a and b as $k \to \infty$ such that $a - a_k = O(1/k^2)$ and
$b - b_k = O(1/k^2)$, then the spectral function ψ of (2.4.5) is continuous within
the interval [a-2b, a+2b] and has a finite number of jumps which are outside or
on the edge of this interval. It should be said that CASE's proofs are rather
obscure. We shall not make use of his results.

MAKI (1976) considers birth and death parameters λ_n and μ_n which are rational
functions of n. For most cases he is able to show that the spectral function is
a step function.

3   DUAL BIRTH-DEATH PROCESSES

3.1   Introduction

By G we shall denote the class of sets of birth-death parameters
$\{\lambda_n, \mu_n: n = 0, 1 \ldots\}$ with $\mu_0 = 0$, and by G^* the class of sets of birth-death
parameters $\{\lambda_n^*, \mu_n^*: n = 0, 1, \ldots\}$ with $\mu_0^* > 0$. The mapping $f: G \to G^*$ defined
by

(3.3.1) $f(\{\lambda_n, \mu_n\}) = \{\lambda_n^*, \mu_n^*\}$

where

(3.1.2) $\lambda_n^* = \mu_{n+1}, \quad \mu_n^* = \lambda_n$ $n = 0, 1, \ldots,$

clearly establishes a $1 - 1$ correspondence between the elements of G and G^*.
Now let $\{\lambda_n, \mu_n\} \in G$ and $\{\lambda_n^*, \mu_n^*\} = f(\{\lambda_n, \mu_n\}) \in G^*$ be two related sets of
birth-death parameters and $\{\pi_n\}$, respectively $\{\pi_n^*\}$, the associated potential
coefficients. The following identities are easily verified in view of (1.3.3) and
(3.1.2).

(3.1.3) $\pi_n^* = \lambda_0 / \lambda_n \pi_n$ $n = 0, 1, \ldots,$

and

(3.1.4) $\pi_n = \mu_0^* / \mu_n^* \pi_n^*$ $n = 0, 1, \ldots.$

Subsequently, with A and A^* denoting the generators that are (uniquely) asso-
ciated with $\{\lambda_n, \mu_n\}$ and $\{\lambda_n^*, \mu_n^*\}$, respectively, it is not difficult to show
that

(3.1.5) $C(A)$ iff $D(A^*)$

and

(3.1.6) $D(A)$ iff $C(A^*)$.

Hence the mapping f remains 1-1 if we restrict G and G^* to those sets of
parameters which satisfy the conditions $C(.)$ and $D(.)$. We have seen in section 1.3
that a set of birth-death parameters which satisfies these conditions uniquely
determines a (natural) birth-death process (or rather, the transition matrix of a
natural process; but since, for the time being, we are not interested in the
initial distribution of a process, we shall identify the process with its transition
matrix). Thus we conclude that (3.1.1) establishes a $1 - 1$ correspondence between

the natural birth-death processes where 0 is a reflecting barrier ($\mu_0 = 0$) and those with an absorbing state -1 ($\mu_0^* > 0$). The corresponding processes were called *dual* to each other by KARLIN and McGREGOR (1957[b]).

3.2   Duality relations

In this section we shall give the relations that exist between, respectively, the transition probability functions, the birth-death polynomials and the spectral functions of dual birth-death processes. Our analysis will be based upon a set $\{\lambda_n, \mu_n\} \in G$ which determines a natural birth-death process with a reflecting barrier 0. All parameters and variables relating to its dual process will be indicated by an asterisk (e.g., A^*, $P^*(t)$, $p_{ij}^*(t)$, $Q_n^*(x)$, $x_{i,n}^*$, $\psi^*(x)$).
In the next theorems U and V denote the infinite matrices

$$
(3.2.1) \qquad U = \begin{pmatrix} 1 & 1 & 1 & 1 & . & . & . \\ 0 & 1 & 1 & 1 & . & . & . \\ 0 & 0 & 1 & 1 & . & . & . \\ . & . & . & . & . & . & . \\ . & . & . & . & . & . & . \end{pmatrix}, \quad V = \begin{pmatrix} 1 & -1 & 0 & 0 & . & . & . \\ 0 & 1 & -1 & 0 & . & . & . \\ 0 & 0 & 1 & -1 & . & . & . \\ . & . & . & . & . & . & . \\ . & . & . & . & . & . & . \end{pmatrix}
$$

that were defined in section 1.2. It is readily seen that the generators A and A^* are related as follows (in this and subsequent statements involving products of infinite matrices, summability requirements to ensure associativity of the multiplication are easily seen to be satisfied, cf. the remark in appendix 2).

$$(3.2.2) \qquad A^* = U^T A^T V^T ; \quad A = U(A^*)^T V.$$

As for the transition matrices $P(.) = (p_{ij}(.))$ and $P^*(.) = (p_{ij}^*(.))$, $i,j = 0, 1, \ldots.$, similar relations hold.

THEOREM 3.2.1   *(i)* $P^*(t) = U^T P^T(t) V^T$; *(ii)* $P(t) = U(P^*(t))^T V.$

PROOF. *(i)*: $D(A^*)$ is supposed to hold whence, by KEMPERMAN's (1962) results, the equations (1.4.8) and (1.4.9) in terms of A^* have a unique solution $P^*(t)$. Therefore, we are done if we can show that $Q^*(t) = U^T P^T(t) V^T$ satisfies the conditions (I) $Q^*(0) = I$ and (II) $(Q^*(t))' = Q^*(t)A^*$. Evidently, (I) holds. Moreover,

$$
\begin{aligned}
(Q^*(t))' &= U^T (P'(t))^T V^T = U^T (P^T(t)A^T)V^T = U^T (P^T(t)(A^T V^T)) = \\
&= U^T (P^T(t)(V^T A^*)) = U^T ((P^T(t)V^T)A^*) = (U^T P^T(t)V^T)A^* = \\
&= Q^*(t)A^*,
\end{aligned}
$$

according to (1.4.7) and (3.2.2). So (II) holds too which proves (i).

(ii): For reasons similar to those above it suffices to show that $Q(t) = U(P^*(t))^T V$ satisfies $Q(0) = I$ and $Q'(t) = Q(t)A$. An additional problem is now that the elements of $Q(t)$ are infinite series, viz.,

$$(Q(t))_{ij} = \sum_{k=i}^{\infty}(p_{jk}^*(t) - p_{j-1,k}^*(t)) \qquad\qquad j > 0$$

and

$$(Q(t))_{i0} = \sum_{k=i}^{\infty}p_{0k}^*(t).$$

It is clear that $Q(t)$ exists. Furthermore, theorem 7 of KARLIN and McGREGOR (1957[a]) implies that

$$\sum_{k=i}^{\infty}(d/dt)p_{jk}^*(t)$$

converges uniformly on every finite interval. Consequently,

$$Q'(t) = (d/dt)(U(P^*(t))^T V) = U((d/dt)P^*(t))^T V.$$

The proof further parallels that of (i).   □

It would seem that part (ii) of theorem 3.2.1 is a simple consequence of part (i). However, the argument

$$P(t) = (UV)P(t)(UV) = U(VP(t)U)V = U(P^*(t))^T V$$

is valid iff (2.1.5) holds, as can be seen by applying the theorem on associativity of products of infinite matrices given in appendix 2. We have still to prove (2.1.5), however, which we shall do now.

Writing out theorem 3.2.1 one has

(3.2.3) $\qquad p_{ij}^*(t) = \sum_{k=0}^{i}(p_{jk}(t) - p_{j+1,k}(t)),$

(3.2.4) $\qquad p_{ij}(t) = \sum_{k=i}^{\infty}(p_{jk}^*(t) - p_{j-1,k}^*(t)) \qquad\qquad j > 0$

and

(3.2.4') $\qquad p_{i0}(t) = \sum_{k=i}^{\infty}p_{0k}^*(t).$

From the above relations it follows at once that $p_{ij}(t) \to 0$ and $p_{ij}^*(t) \to 0$ as $i \to \infty$, i.e., (2.1.5) holds.

Before giving the relations between the birth-death polynomials of dual birth-death processes, we remark that the recurrence relations (2.2.1) can be written more compactly as

(3.2.5) $Q_0(x) = 1$; $-xQ(x) = AQ(x)$,

respectively,

(3.2.6) $Q_0^*(x) = 1$; $-xQ^*(x) = A^* Q^*(x)$,

where

(3.2.7) $Q(x) = (Q_0(x), Q_1(x), \ldots.)^T$; $Q^*(x) = (Q_0^*(x), Q_1^*(x), \ldots.)^T$.

Also we note that with

(3.2.8) $\Pi = \text{diag}(\pi_0, \pi_1, \ldots.)$; $\Pi^* = \text{diag}(\pi_0^*, \pi_1^*, \ldots.)$

one has

(3.2.9) $\Pi A = A^T \Pi$; $\Pi^* A^* = (A^*)^T \Pi^*$,

which is straightforwardly verified.

THEOREM 3.2.2. *(i)* $\pi_{n+1} Q_{n+1}(x) = Q_{n+1}^*(x) - Q_n^*(x)$;

  (ii) $-x Q_n^*(x) = \lambda_n \pi_n (Q_{n+1}(x) - Q_n(x))$.

PROOF. *(i)*: Let $R_0(x) = 1 = Q_0^*(x)/\pi_0$,

  $R_{n+1}(x) = (Q_{n+1}^*(x) - Q_n^*(x))/\pi_{n+1}$ $n \geq 0$,

and $R(x) = (R_0(x), R_1(x), \ldots.)^T$. Then

  $R(x) = \Pi^{-1} V^T Q^*(x)$.

Now considering (3.2.2) and (3.2.9) one obtains from (3.2.6)

  $-x R(x) = \Pi^{-1} V^T (-x Q(x)) = \Pi^{-1} V^T (A^* Q^*(x)) = \Pi^{-1}((A^T V^T) Q^*(x)) =$

  $= \Pi^{-1}(A^T(V^T Q^*(x))) = (\Pi^{-1} A^T)(V^T Q^*(x)) = (A \Pi^{-1})(V^T Q^*(x)) =$

  $= A(\Pi^{-1} V^T Q^*(x)) = A R(x)$.

Hence $R(x) = Q(x)$.

  (ii): Similarly.   □

The polynomials $Q_n(x)$ are orthogonal with respect to the spectral function ψ, while the polynomials $Q_n^*(x)$ are orthogonal with respect to the spectral function ψ^*. The relation between ψ and ψ^* was given by KARLIN and McGREGOR (1957^a), lemmas 2 and 3, as follows (in their terminology $Q_n^*(x) = H_{n+1}(x)/-x$).

THEOREM 3.2.3. (i) $\int_0^x d\psi^*(\xi) = \int_0^x \xi d\psi(\xi)/\lambda_0$ $\qquad\qquad x \geq 0.$

$\qquad\qquad (ii)$ $\Delta\psi(0) = 1 - \lambda_0 \int_{+0}^{\infty} d\psi^*(\xi)/\xi,$

$\qquad\qquad\qquad \int_0^x d\psi(\xi) = \Delta\psi(0) + \lambda_0 \int_{+0}^x d\psi^*(\xi)/\xi$ $\qquad\qquad x > 0.$

3.3 Ergodic properties

We conclude this chapter with a description of the behaviour of the dual birth-death processes as t approaches infinity, and some related facts. According to KARLIN and McGREGOR (1957^b) the process with generator A (i.e., with reflecting barrier 0) is

(3.3.1)   transient   iff $\Sigma(1/\lambda_n\pi_n) < \infty$

(3.3.2)   null recurrent   iff $\Sigma(1/\lambda_n\pi_n) = \Sigma\pi_n = \infty$

(3.3.3)   positive recurrent iff $\Sigma\pi_n < \infty.$

We recall that

(3.3.4) $\Sigma(\pi_n + 1/\lambda_n\pi_n) = \infty,$

as a consequence of the fact that C(A) and D(A) are satisfied.
The classical ergodic theorem concerning the behaviour of $p_{ij}(t)$ as $t \to \infty$ states that

(3.3.5) $\lim_{t\to\infty} p_{ij}(t) = \pi_j/\Sigma\pi_n ,$

which is to be interpreted as zero if $\Sigma\pi_n = \infty$, i.e., if the process is transient or null recurrent.
It is seen from (3.2.3) and (3.3.5) that

(3.3.6) $\lim_{t\to\infty} p_{ij}^*(t) = 0.$

Note, however, that (3.2.4) and (3.2.4') imply

(3.3.7) $\sum_{j=0}^{\infty} p_{ij}^*(t) = \sum_{j=0}^{i} p_{0j}(t) ,$

so that

(3.3.8) $\qquad \lim\limits_{t\to\infty} \sum\limits_{j=0}^{\infty} p_{ij}^{*}(t) = \sum\limits_{n=0}^{i}\pi_n / \sum\limits_{n=0}^{\infty}\pi_n$.

From (1.4.13) and (3.3.8) we conclude that absorption at -1 will not take place with a positive probability if $\Sigma\pi_n < \infty$: the absorption is called *transient*. It is called *certain* if the probability of eventual absorption at -1 is one, i.e., $\lim\limits_{t\to\infty} \sum\limits_{j=0}^{\infty} p_{ij}^{*}(t) = 0$, which is equivalent to $\Sigma\pi_n = \infty$.

For the reflecting barrier process CALLAERT (1971, 1974) has given the links that exist between the positions of the limit points x_1 and x_2 on the one hand and the criteria (3.3.1) - (3.3.3) on the other. His results are

(3.3.9) $\qquad x_1 > 0 \qquad\qquad\qquad$ iff $\Sigma(1/\lambda_n\pi_n) < \infty$

(3.3.10) $\qquad x_1 = x_2 = 0 \qquad\qquad$ iff $\Sigma\pi_n = \Sigma(1/\lambda_n\pi_n) = \infty$

(3.3.11) $\qquad x_1 = 0,\ x_2 > 0 \qquad$ iff $\Sigma\pi_n < \infty$.

The above results together with theorem 3.2.3 *(i)* and the theorems 2.2.6 and 2.2.7 are readily seen to imply the following

(3.3.12) $\qquad x_i^{*} = x_1 > 0 \quad$ for all i $\qquad$ iff $\Sigma(1/\lambda_n\pi_n) < \infty$

(3.3.13) $\qquad x_i^{*} = x_i = 0 \quad$ for all i $\qquad$ iff $\Sigma\pi_n = \Sigma(1/\lambda_n\pi_n) = \infty$

(3.3.14) $\qquad x_1 = 0 < x_{i+1} = x_i^{*} \quad$ for all i $\quad$ iff $\Sigma\pi_n < \infty$.

From these results, (3.1.3) and (3.1.4) we can infer the validity of the next assertions which supplement CALLAERT's results (3.3.9) - (3.3.11).

(3.3.15) $\qquad x_1^{*} > 0 \qquad$ iff $\Sigma\pi_n^{*} < \infty \quad$ or $\quad \Sigma(1/\lambda_n^{*}\pi_n^{*}) < \infty$

(3.3.16) $\qquad x_1^{*} = x_2^{*} = 0 \quad$ iff $\Sigma\pi_n^{*} = \Sigma(1/\lambda_n^{*}\pi_n^{*}) = \infty$.

4.   STOCHASTIC MONOTONICITY: GENERAL RESULTS

4.1   The case $\mu_0 = 0$
——————————————————

Let $\{\lambda_n, \mu_n: n = 0, 1,.....\}$, with $\mu_0 = 0$, be the set of parameters of a
natural birth-death process $\{X(t): 0 \leq t < \infty\}$ where 0 is a reflecting barrier.
The initial distribution vector of $\{X(t)\}$ will be denoted by $\underline{q} = (q_0, q_1,....)^T$,
i.e.,

$$q_i = p_i(0) = Pr\{X(0) = i\} \qquad\qquad i = 0, 1, ...;$$

otherwise the notation of section 1.4 will be used. We have

(4.1.1) $\underline{q} \geq \underline{0}, \ \underline{q}^T \underline{1} = 1,$

where vector inequality is defined by (1.2.15) and $\underline{0}$ and $\underline{1}$ are the column
vectors consisting of 0's and 1's, respectively. We recall that

(4.1.2) $\underline{p}^T(t) = \underline{q}^T P(t),$

where $\underline{p}^T(t) = (p_0(t), p_1(t),.....)$ with $p_i(t) = Pr\{X(t) = i\}$ and $P(t)$ the transition matrix of $\{X(t)\}$. More generally one has

(4.1.3) $\underline{p}^T(t+s) = \underline{p}^T(t)P(s)$

From (1.4.14) and (4.1.1) it follows that for all $t \geq 0$

(4.1.4) $\underline{p}^T(t)\underline{1} = 1,$

so that $\underline{p}(t)$ is indeed a probability distribution vector.
In what follows a fundamental role is played by the matrix $E(t) = (e_{ij}(t))$, $i,j = 0,1,...$, which is defined for $t \geq 0$ by

(4.1.5) $E(t) \equiv P(t)AU,$

where A is the generator of $\{X(t)\}$ and U the matrix given by (1.2.10). It is
readily seen that

(4.1.6) $E(0) = AU = \begin{pmatrix} -\lambda_0 & 0 & 0 & . & . & . & . & . \\ \mu_1 & \lambda_1 & 0 & . & . & . & . & . \\ 0 & \mu_2 & -\lambda_2 & . & . & . & . & . \\ . & . & . & . & . & . & . & . \\ . & . & . & . & . & . & . & . \end{pmatrix} \ .$

In the next lemma $P^*(.)$ denotes the transition matrix which is dual to $P(.)$ in the sense of chapter 3.

LEMMA 4.1.1. $E(t) = AU(P^*(t))^T$ *for all* $t \geq 0$.

PROOF. From (1.4.7), (1.4.8) and (4.1.5) it is seen that

$$E(t) = (P(t)A)U = (AP(t))U = AP(t)U .$$

Substitution of $P(t) = U(P^*(t))^T V$ (theorem 3.2.1) yields the desired result in view of (1.2.12). □

The vector $\underline{e}(t) = (e_0(t), e_1(t), \ldots)^T$ is defined for $t \geq 0$ as

$$(4.1.7) \qquad \underline{e}^T(t) \equiv (d/dt)(\underline{p}^T(t)U).$$

We note that for all i the series

$$p_i(t) = \sum_k q_k p_{ki}(t)$$

converges uniformly, whence by (1.3.9) for all i the series

$$\sum_k q_k p'_{ki}(t)$$

converges uniformly. Consequently, $p_i(t)$ has for all i a finite derivative. Hence $\underline{e}(t)$ is well defined and has finite components for all $t \geq 0$. Moreover,

$$(4.1.8) \qquad \underline{e}^T(t) = \underline{q}^T P'(t)U .$$

From (1.4.8) one obtains

$$\underline{q}^T P'(t)U = \underline{q}^T(P(t)A)U = \underline{q}^T P(t)AU,$$

so that with (4.1.2)

$$(4.1.9) \qquad \underline{e}^T(t) = \underline{p}^T(t)AU,$$

and with (4.1.5)

$$(4.1.10) \qquad \underline{e}^T(t) = \underline{q}^T E(t).$$

The next lemma is a key result. The $\{\pi_n\}$ are the potential coefficients of the birth-death process $\{X(t)\}$.

LEMMA 4.1.2. *If the sequence* $(q_n/\pi_n)_n$ *is bounded, then for* s, t $\geq$ 0,

$$\underline{e}(t+s) = P^*(s)\underline{e}(t) \ .$$

PROOF. We obtain from (4.1.9), (4.1.3) and (4.1.5)

$$\underline{e}^T(t+s) = \underline{p}^T(t+s)AU = (\underline{p}^T(t)P(s))AU = \underline{p}^T(t)(P(s)AU) =$$

$$= \underline{p}^T(t)E(s) \ .$$

Hence, by lemma 4.1.1,

$$\underline{e}^T(t+s) = \underline{p}^T(t)(AU(P^*(s))^T) \ .$$

The lemma now follows at once from (4.1.9) if we can prove that

$$(4.1.11) \qquad \underline{p}^T(t)(AU(P^*(s))^T) = (\underline{p}^T(t)AU)(P^*(s))^T \ .$$

To check whether (4.1.11) holds, we apply theorem A.2.2 of appendix 2. After some afgebra one sees that (4.1.11) is valid iff for all j

$$(4.1.12) \qquad -\lambda_m p_m(t)p_{jm}^*(s) \to 0 \text{ as } m \to \infty \ .$$

With (2.1.6) and (3.1.3) one gets

$$\lambda_m p_m(t)p_{jm}^*(s) = \lambda_0 p_m(t)p_{mj}^*(s)/\pi_m\pi_j^* \ .$$

Considering that $p_{mj}^*(s) \to 0$ as $m \to \infty$ (see (2.1.5)) it is therefore sufficient that $p_m(t)/\pi_m$ is bounded as $m \to \infty$ for (4.1.12) to hold. Using (2.1.6) again one obtains with (4.1.2)

$$p_m(t)/\pi_m = \sum_k q_k p_{km}(t)/\pi_m = \sum_k q_k p_{mk}(t)/\pi_k \ .$$

Hence, if $q_k/\pi_k < M < \infty$, then

$$p_m(t)/\pi_m < M\sum_k p_{mk}(t) = M < \infty \ . \qquad\qquad \square$$

The matrix E(t) and the vector $\underline{e}(t)$ have been introduced because of their importance when investigating the question whether the birth-death process {X(t)} is stochastically monotone on some interval or not. The next theorem, which is the main application in this section of the preceding lemma, gives evidence of this fact.

THEOREM 4.1.3. *Let the sequence* $(q_n/\pi_n)_n$ *be bounded. The natural birth-death process* $\{X(t): 0 \leq t < \infty\}$ *is strictly stochastically increasing (decreasing) on the interval* (t_1, ∞) *iff*

(4.1.13) $\underline{e}(t_1) \leq \underline{0} \; (\underline{e}(t_1) \geq \underline{0})$ *and* $\underline{e}(t_1) \neq \underline{0}$.

PROOF. If $\{X(t)\}$ is strictly stochastically increasing (decreasing) on the interval (t_1, ∞), then, by (1.2.20),

(4.1.14) $\underline{p}^T(\tau_2)U < \underline{p}^T(\tau_1)U \; (\underline{p}^T(\tau_1)U < \underline{p}^T(\tau_2))$

for every pair τ_1, τ_2 with $0 \leq t_1 \leq \tau_1 < \tau_2 < \infty$, so that in particular for all $\tau > t_1$

$$\underline{p}^T(\tau)U < \underline{p}^T(t_1)U \; (\underline{p}^T(t_1)U < \underline{p}^T(\tau)U).$$

Consequently, by (4.1.7),

$$\underline{e}(t_1) \leq \underline{0} \; (\underline{e}(t_1) \geq \underline{0}).$$

Moreover, in view of lemma 4.1.2, $\underline{e}(t_1) = \underline{0}$ implies $\underline{e}(t) = \underline{0}$ for all $t \geq t_1$, which is clearly a contradiction.
On the other hand, if $\underline{e}(t_1) \leq \underline{0} \; (\underline{e}(t_1) \geq \underline{0})$ and $\underline{e}(t_1) \neq \underline{0}$, then, because of (2.1.2) and lemma 4.1.2, $\underline{e}(t_1+s) < \underline{0} \; (\underline{e}(t_1+s) > \underline{0})$ for all $s > 0$. Consequently, (4.1.14) holds true for every pair τ_1, τ_2 with $t_1 \leq \tau_1 < \tau_2 < \infty$, whence $\{X(t)\}$ is strictly stochastically increasing (decreasing) on the interval (t_1, ∞). $\square$

It is easily verified that if we leave out the word *strictly* in theorem 4.1.3, then the condition (4.1.13) should become

(4.1.13') $\underline{e}(t_1) \leq \underline{0} \; (\underline{e}(t_1) \geq \underline{0})$,

in order that a valid theorem remains. Thus we see that if the process $\{X(t)\}$ is stochastically monotone on (t_1, ∞), then it is either strictly stochastically monotone on (t_1, ∞) or it has $\underline{e}(t_1) = \underline{0}$, the latter implying, by lemma 4.1.2, $\underline{e}(t) = \underline{0}$ for all $t \geq 0$, whence

(4.1.15) $p_i(t) = q_i$ for all i and $t \geq 0$.

It will be clear now that we shall be interested in strict monotonicity only in future.
In this context we remark that a somewhat stronger version of theorem 1.3.1 is implicit in the proof of the above theorem, in that the word *strictly*

may be inserted before stochastically.

The necessary and sufficient condition for $\{X(t)\}$ to be stochastically increasing on the interval $(0,\infty)$, as given by KEILSON and KESTER (1977), can be obtained from theorem 4.1.3 in a straightforward manner. Namely, this theorem states that fqr $\{X(t)\}$ to be strictly stochastically increasing on $(0,\infty)$ it is necessary and sufficient that

(4.1.16) $\underline{e}(0) \leq \underline{0}$ and $\underline{e}(0) \neq \underline{0}$.

From (4.1.6) and (4.1.10) one has

$$e_i(0) = (\underline{q}^T E(0))_i = \mu_{i+1} q_{i+1} - \lambda_i q_i.$$

Since $\lambda_i \pi_i = \mu_{i+1} \pi_{i+1}$ it follows that (4.1.16) holds iff the sequence $(q_n/\pi_n)_n$ is non-increasing and non-constant, which is KEILSON and KESTER's (1977) result except that they allow the sequence $(q_n/\pi_n)_n$ to be constant (which results in a process with constant absolute probabilities).

Similarly one can show that $\{X(t)\}$ is strictly stochastically decreasing on $(0,\infty)$ iff the sequence $(q_n/\pi_n)_n$ is non-decreasing and non-constant (it should be bounded though). Summarizing we have the next theorem.

THEOREM 4.1.4 *(i)* (KEILSON and KESTER (1977)). *The natural birth-death process $\{X(t): 0 \leq t < \infty\}$ is strictly stochastically increasing on $(0,\infty)$ iff the sequence $(q_n/\pi_n)_n$ is non-increasing and $q_0 \neq q_n/\pi_n$ for some n.*

(ii) Let the sequence $(q_n/\pi_n)_n$ be bounded. The process $\{X(t)\}$ is strictly stochastically decreasing on $(0,\infty)$ iff $(q_n/\pi_n)_n$ is non-decreasing and $q_0 \neq q_n/\pi_n$ for some n.

4.2   The case $\mu_0 > 0$

Let $\{\lambda_n^*, \mu_n^*\}$, with $\mu_0^* > 0$, be the set of parameters of a natural birth-death process $\{X^*(t): 0 \leq t < \infty\}$ with absorbing state -1. As seen in chapter 3 we may consider this process as dual to a natural birth-death process with parameters $\{\lambda_n, \mu_n\}$, where $\mu_0 = 0$, $\lambda_n = \mu_n^*$ for $n \geq 0$ and $\mu_n = \lambda_{n-1}^*$ for $n \geq 1$. As usual an asterisk distinguishes the parameters and variables of the process $\{X^*(t)\}$ from the corresponding quantities of its dual process $\{X(t)\}$. We let $\underline{q}^* = (q_{-1}^*, q_0^*, q_1^*, \ldots)^T$ be the initial distribution vector of $\{X^*(t)\}$ and use the notation $\underline{p}^*(t) = (p_{-1}^*(t), p_0^*(t), p_1^*(t), \ldots)^T$, where $p_i^*(t) = \Pr\{X^*(t) = i\}$. The vector $\underline{e}^*(t) = (e_{-1}^*(t), e_0^*(t), e_1^*(t), \ldots)^T$ is defined for $t \geq 0$ as

(4.2.1) $(\underline{e}^*(t))^T \equiv (d/dt)(\underline{p}^*(t))^T U)$.

By an argument similar to that in the previous section one can show that $p_i^*(t)$ has a finite derivative for $i = -1, 0, 1, \ldots$, so that $\underline{e}^*(t)$ is well-defined. In the next lemma $(\underline{0}\ E(t))$ denotes the matrix defined by (4.1.5) preceded by a column of zeros.

LEMMA 4.2.1. $\underline{e}^*(t) = -(\underline{0}\ E(t))\underline{q}^*$ *for all* $t \geq 0$.

PROOF. Let us denote by $\overline{P}^*(t)$ the complete matrix of transition probabilities of $\{X^*(t)\}$, i.e.,

$$(4.2.2) \qquad \overline{P}^*(t) = \left(\begin{array}{c|ccccc} 1 & 0 & 0 & 0 & 0 & \cdots \\ \hline p_{0,-1}^*(t) & & & & & \\ p_{1,-1}^*(t) & & & P^*(t) & & \\ \vdots & & & & & \end{array} \right),$$

where $P^*(t)$ is the transition matrix of $\{X^*(t)\}$. The forward equations (1.3.9) can now be expressed as

(4.2.3) $(d/dt)\overline{P}^*(t) = \overline{P}^*(t)\overline{A}^*$,

where $\overline{A}^*$ is the q-matrix of $\{X^*(t)\}$, i.e.,

$$(4.2.4) \quad \overline{A}^* = \left(\begin{array}{c|ccc} 0 & 0 & 0 & \cdots \\ \hline \mu_0^* & & & \\ 0 & & A^* & \\ \vdots & & & \end{array} \right) = \left(\begin{array}{cccc} 0 & 0 & 0 & \cdots \\ \mu_0^* & -(\lambda_0^*+\mu_0^*) & \lambda_0^* & \cdots \\ 0 & \mu_1^* & -(\lambda_1^*+\mu_1^*) & \cdots \\ \cdots & \cdots & \cdots & \cdots \\ \cdots & \cdots & \cdots & \cdots \end{array} \right).$$

Obviously,

(4.2.5) $(\underline{p}^*(t))^T = (\underline{q}^*)^T \overline{P}^*(t)$,

whence (4.2.1) becomes

(4.2.6) $(\underline{e}^*(t))^T = (d/dt)((\underline{q}^*)^T \overline{P}^*(t)U) = (\underline{q}^*)^T ((d/dt)\overline{P}^*(t))U =$
               $= (\underline{q}^*)^T (\overline{P}^*(t)\overline{A}^*)U = (\underline{q}^*)^T \overline{P}^*(t)(\overline{A}^* U)$.

Some simple algebra shows that

(4.2.7) $\overline{A}^* U = -(\underline{0}\ AU)^T$,

where A is the generator of $\{X(t)\}$, and subsequently that

$$\overline{P}^*(t)(\underline{0}\ AU)^T = (\underline{0}\ AU(P^*(t))^T)^T = (\underline{0}\ E(t))^T ,$$

according to lemma 4.1.1. The lemma follows.   □

The next lemma is the analogue of lemma 4.1.2.

LEMMA 4.2.2. *If the sequence* $(q_n^*/\pi_n^*: n = 0, 1, \ldots\ldots)$ *is bounded, then for* $s,t \geq 0$

$$\underline{e}^*(t+s) = P(s)\underline{e}^*(t) .$$

PROOF. From (4.2.5), (4.2.6) and (4.2.7) it follows that

(4.2.8) $\underline{e}^*(t) = -(\underline{0}\ AU)\underline{p}^*(t)$,

so that in particular

$$\underline{e}^*(t+s) = -(\underline{0}\ AU)\underline{p}^*(t+s) .$$

Now, by (1.1.7) and (4.2.5)

$$(\underline{p}^*(t+s))^T = (\underline{q}^*)^T \overline{P}^*(t+s) = (\underline{q}^*)^T \overline{P}^*(t)\overline{P}^*(s) = (\underline{p}^*(t))^T \overline{P}^*(s),$$

whence

$$\underline{e}^*(t+s) = -(\underline{0}\ AU)((\overline{P}^*(s))^T \underline{p}^*(t)) = -((\underline{0}\ AU)(\overline{P}^*(s))^T)\underline{p}^*(t) =$$
$$= -(\underline{0}\ AU(P^*(s))^T)\underline{p}^*(t) ,$$

as can easily be verified. By lemma 4.1.1 and (4.1.5) we now have

$$\underline{e}^*(t+s) = -(\underline{0}\ E(s))\underline{p}^*(t) = -(\underline{0}\ P(s)AU)\underline{p}^*(t) =$$
$$= -(P(s)(\underline{0}\ AU))\underline{p}^*(t) .$$

The lemma follows if we can show that

$$(P(s)(\underline{0}\ AU))\underline{p}^*(t) = P(s)((\underline{0}\ AU)\underline{p}^*(t)),$$

given that the sequence $(q_n^*/\pi_n^*: n = 0, 1, \ldots\ldots)$ is bounded. This can be done by methods similar to those in the proof of lemma 4.1.2.   □

It is obvious that the birth-death process $\{X^*(t)\}$ cannot be strictly stochastically increasing, since $p_{-1}^*(t)$, the first component of $(\underline{p}^*(t))^T U$, cannot decrease with time, -1 being an absorbing state. We even have by

(1.4.6) and (2.1.2)

(4.2.9) $(d/dt)p_{-1}^{*}(t) = \mu_0^{*} \sum_{k=0}^{\infty} q_k^{*} p_{k0}^{*}(t) > 0$

for $t > 0$, provided $q_{-1}^{*} < 1$. Thus $\{X^{*}(t)\}$ cannot be stochastically increasing
if $q_{-1}^{*} < 1$. However, the following holds and makes sense.

THEOREM 4.2.3. *Let the sequence* $(q_n^{*}/\pi_n^{*}: n = 0, 1, \ldots\ldots)$ *be bounded. The*
natural birth-death process $\{X^{*}(t): 0 \le t < \infty\}$ *is strictly stochastically decreasing*
on the interval (t_1, ∞) *iff*

(4.2.10) $\underline{e}^{*}(t_1) \ge \underline{0}$ *and* $\underline{e}^{*}(t_1) \neq \underline{0}$.

The proof of the above theorem is analogous to the proof of theorem 4.1.3.
Further, it is easy to see that if $\{X^{*}(t)\}$ is stochastically decreasing, then
either it is strictly stochastically decreasing, or it has $p_i^{*}(t) = \delta_{i,-1}$
for all $t \ge 0$.
For $\{X^{*}(t)\}$ to be strictly stochastically decreasing on $(0,\infty)$ it is,
according to the above theorem, necessary and sufficient that $\underline{e}^{*}(0) \ge \underline{0}$ and
$\underline{e}^{*}(0) \neq \underline{0}$. From lemma 4.2.1 and (4.1.6) one sees that

$\underline{e}^{*}(0) = -(\underline{0} \ AU)\underline{q}^{*}$.

The next theorem is now easily verified.

THEOREM 4.2.4. *Let the sequence* $(q_n^{*}/\pi_n^{*}: n = 0, 1, \ldots\ldots)$ *be bounded. The*
natural birth-death process $\{X^{*}(t): 0 \le t < \infty\}$ *is strictly stochastically decreasing*
on $(0,\infty)$ *iff* $(q_n^{*}/\pi_n^{*})_n$ *is non-decreasing and* $q_0^{*} \neq q_n^{*}/\pi_n^{*}$ *for some n.*

4.3   Properties of E(t)

We are interested in the problem whether a natural birth-death process with an
arbitrary initial distribution is strictly stochastically monotone on some
interval or not. In view of the theorems 4.1.3 and 4.2.3 this is the case when
the associated vector $\underline{e}(t)$ (or $\underline{e}^{*}(t)$) has the properties $\underline{e}(t) \le \underline{0}$ ($\underline{e}(t) \ge \underline{0}$)
and $\underline{e}(t) \neq \underline{0}$ for some value of t. It will appear in the next chapter that the
question whether $\underline{e}(t)$ satisfies this condition can be answered completely if the
initial distribution is such that it assigns unit probability to a single state.
From (4.1.10) and lemma 4.2.1 we see that in this case the associated vectors are
in fact rows, respectively columns, of the matrix E(t).
It should be borne in mind that, as in chapter 3, a set of birth-death parameters

$\{\lambda_n, \mu_n\}$ with $\mu_0 = 0$, is the basis of our considerations. This set corresponds directly to the reflecting barrier process considered in section 4.1, whereas it is the dual set of the absorbing process considered in section 4.2. In this manner the results of the preceding two sections are centred round the same matrix $E(t)$.

From (4.1.5) and (4.1.6) one obtains

(4.3.1) $\qquad e_{ij}(t) = \mu_{j+1} p_{i,j+1}(t) - \lambda_j p_{ij}(t)$,

and from lemma 4.1.1 it is seen that also

(4.3.2) $\qquad \begin{aligned} &e_{0j}(t) = -\lambda_0 p_{j0}^*(t) \\ &e_{ij}(t) = -\lambda_i p_{ji}^*(t) + \mu_i p_{j,i-1}^*(t) \end{aligned}$ $\qquad\qquad\qquad\qquad$ $i = 1, 2, \ldots$.

As a consequence of (4.3.1) and (2.1.5)

(4.3.3) $\qquad e_{ij}(t) \to 0$ as $i \to \infty$,

and as a consequence of (4.3.2) and (2.1.5)

(4.3.4) $\qquad e_{ij}(t) \to 0$ as $j \to \infty$.

Moreover, (4.3.1), (3.3.5) and the fact that $\lambda_n \pi_n = \mu_{n+1} \pi_{n+1}$ show that

(4.3.5) $\qquad e_{ij}(t) \to 0$ as $t \to \infty$.

LEMMA 4.3.1. *Let* $t > 0$.

(i) *For all* $i, j = 0, 1, \ldots\ldots, a$ $k \geq j$ *exists such that* $e_{ik}(t) < 0$.
(ii) *For all* $i, j = 0, 1, \ldots\ldots, a$ $k \geq i$ *exists such that* $e_{kj}(t) > 0$.

PROOF. *(i):* From (4.3.1) and (2.1.6) one has

$$e_{ik}(t) = (\mu_{k+1} \pi_{k+1} p_{k+1,i}(t) - \lambda_k \pi_k p_{ki}(t))/\pi_i ,$$

which yields

(4.3.6) $\qquad e_{ik}(t) = \lambda_k \pi_k (p_{k+1,i}(t) - p_{ki}(t))/\pi_i$.

Let i,j be fixed. If $e_{ik}(t) \geq 0$ for all $k \geq j$, then $p_{k+1,i}(t) \geq p_{ki}(t)$ for all $k \geq j$, by (4.3.6). This, however, contradicts (2.1.2) and (2.1.5).

$\qquad$ *(ii):* Similarly with (4.3.2) and (2.1.6). $\qquad\qquad\qquad\qquad\qquad\qquad\qquad\qquad$ $\square$

As a consequence of the above lemma the next two theorems hold, where the terminology of section 4.1, respectively 4.2, is used.

THEOREM 4.3.2. *If* $q_n = \delta_{in}$, $n = 0, 1, \ldots$, *then the natural birth-death process* $\{X(t): 0 \leq t < \infty\}$ *is not stochastically decreasing on any interval.*

PROOF. If $q_n = \delta_{in}$, then, by (4.1.10), $\underline{e}(t)$ is the ith row of $E(t)$. According to lemma 4.3.1 *(i)* this row contains negative components for all $t > 0$, and, by (4.1.6), also for $t = 0$. Theorem 4.1.3 now implies that the process cannot be strictly stochastically decreasing on any interval. Further, we can omit the word *strictly*, considering the remarks after theorem 4.1.3 and the fact that $\underline{e}(t) \neq \underline{0}$. $\square$

THEOREM 4.3.3. *If* $q_n^* = \delta_{jn}$, $n = 0, 1, \ldots$, *then the natural birth-death process* $\{X^*(t): 0 \leq t < \infty\}$ *is not stochastically decreasing on any interval.*

PROOF. If $q_n^* = \delta_{jn}$ with $j \geq 0$, then, by lemma 4.2.1, $-\underline{e}^*(t)$ is the jth column of $E(t)$. According to lemma 4.3.1 *(ii)* this column contains positive components for all $t > 0$, and, by (4.1.6), also for $t = 0$. The result now follows from theorem 4.2.3, considering that monotonicity implies strict monotonicity if $q_{-1}^* = 0$. $\square$

In view of the above theorem we restrict ourselves to the case $\mu_0 = 0$ in the first three sections of the next chapter, where we analyse the situation that the process starts in a fixed state. Considering theorem 4.3.2 the problem is then whether the process is strictly stochastically *increasing* on some interval or not.

5. STOCHASTIC MONOTONICITY: DEPENDENCE ON THE INITIAL STATE DISTRIBUTION

5.1 Introduction to the case of fixed initial state

In the first three sections of this chapter we consider a natural birth-death
process for which 0 is a reflecting barrier with initial distribution vector
$\underline{q} = (q_0, q_1, \ldots)^T$, where $q_n = \delta_{in}$ for some fixed i and all n. The process is
denoted by $\{X_i(t)\} = \{X_i(t) : 0 \le t < \infty\}$ and the corresponding time dependent vector
defined by (4.1.7), which, according to (4.1.10), is the ith row of the matrix
$E(t) = (e_{ij}(t))$, is denoted by $\underline{e}_i(t)$, i.e.,

$$\underline{e}_i(t) = (e_{i0}(t), e_{i1}(t), \ldots)^T .$$

To determine whether $\{X_i(t)\}$ is strictly stochastically increasing on an
interval we argue as follows. Considering that $q_n = \delta_{in}$, the sequence $(q_n/\pi_n)_n$
is bounded, whence lemma 4.1.2 yields

(5.1.1) $\underline{e}_i(t) = P^*(t)\underline{e}_i(0),$

where $P^*(t)$ is the transition matrix of the dual process (as usual an asterisk
refers to the dual process). Theorem 2.1.1 therefore implies that for $t > 0$

(5.1.2) $S^+(\underline{e}_i(t)) \le S^-(\underline{e}_i(0)) .$

From (4.1.6) we see that $\underline{e}_i(0)$ is the ith row of the matrix AU, whence

(5.1.3) $S^-(\underline{e}_0(0)) = 0$

and

(5.1.4) $S^-(\underline{e}_i(0)) = 1$ $i = 1, 2, \ldots$.

Combining (5.1.2) – (5.1.4) it emerges that for $t > 0$

(5.1.5) $S^+(\underline{e}_0(t)) = 0$

and

(5.1.6) $S^+(\underline{e}_i(t)) \le 1$ $i = 1, 2, \ldots$.

Before proceeding with the most interesting case $i > 0$ we remark that $\underline{e}_0(0)$
satisfies the conditions of theorem 4.1.3, so that we have the next theorem.

THEOREM 5.1.1. *The natural birth-death process $\{X_0(t)\}$ is strictly stochastically increasing on $(0,\infty)$.*

The above theorem is well known and in fact incorporated in KEILSON and KESTER's (1977) result which is given in theorem 4.1.4 *(i)*.
For $i > 0$ it is evident that $\{X_i(t)\}$ is not strictly stochastically increasing on an interval of the form $(0, t_1)$. However, lemma 4.3.1 *(i)* and (5.1.6) are easily seen to imply the next lemma.

LEMMA 5.1.2. *Let $i > 0$ and $t > 0$. Then $\underline{e}_i(t) \leq \underline{0}$ and $\underline{e}_i(t) \neq \underline{0}$ iff $e_{i0}(t) \leq 0$.*

Combining this lemma and theorem 4.1.3 we obtain the following important result.

THEOREM 5.1.3. *The natural birth-death process $\{X_i(t)\}$, with $i > 0$, is strictly stochastically increasing on the interval (t_1, ∞), with $t_1 > 0$, iff $e_{i0}(t_1) \leq 0$.*

REMARK 5.1.4. From lemma 5.1.2 one sees that $e_{i0}(t_1) = 0$, with $t_1 > 0$, implies $\underline{e}_i(t_1) \leq \underline{0}$ and $\underline{e}_i(t_1) \neq \underline{0}$, which, considering lemma 4.1.2 and (2.1.2), implies $\underline{e}_i(t_1 + s) < \underline{0}$ for all $s > 0$, whence in particular $e_{i0}(t_1 + s) < 0$ for all $s > 0$. Thus if the equation

(5.1.7) $e_{i0}(t) = 0$

has a solution for $t > 0$, $i > 0$, then it has a unique solution τ_i, say. Moreover, from theorem 5.1.3 we conclude that $\{X_i(t)\}$ is strictly stochastically increasing precisely on the interval (τ_i, ∞). It is interesting to mention that $\tau_{i+1} > \tau_i$ (provided these quantities exist). This statement can be proven by performing a sign variation analysis on the first column of the matrix $E(t)$, on the basis of theorem 2.1.1 and the lemmas 4.2.1 and 4.2.2.

In the next two sections we will determine the sign of $e_{i0}(t)$ as t approaches infinity in the two cases $\Sigma\pi_n = \infty$ and $\Sigma\pi_n < \infty$, respectively. Once we know this sign we also know whether $\{X_i(t)\}$ is stochastically increasing in the long run (if the sign is negative) or not (if the sign is positive).
We note that $e_{i0}(t) \to 0$ as $t \to \infty$, according to (4.3.5), which gives us no information. But the following representation will be very useful.

LEMMA 5.1.5. $e_{ij}(t) = -\int_0^\infty \exp(-xt) Q_i(x) Q_j^*(x) x \, d\psi(x)$.

PROOF. Substitution of (2.2.3) in (4.3.1) yields the desired result in view of theorem 3.2.2 *(ii)* and the fact that $\lambda_j\pi_j = \mu_{j+1}\pi_{j+1}$.   □

Since $Q_0^*(x) = 1$, the above lemma implies

$$(5.1.8) \qquad e_{i0}(t) = -\int_0^\infty \exp(-xt)Q_i(x)xd\psi(x) ,$$

for which we may write, in view of theorem 2.2.6,

$$(5.1.9) \qquad e_{i0}(t) = -\int_{x_1}^\infty \exp(-xt)Q_i(x)xd\psi(x) .$$

5.2 The transient and null recurrent process

By (3.3.1) – (3.3.4) we know that the process $\{X_i(t)\}$ is transient or null recurrent iff $\Sigma\pi_n = \infty$.

LEMMA 5.2.1. *Let* $i > 0$. *Iff* $\Sigma\pi_n = \infty$, *then* $e_{i0}(t) < 0$ *for t sufficiently large.*

PROOF. In view of theorem 2.2.6 and (5.1.8) the results of appendix 3 apply with $F \equiv \psi$, $u_k \equiv x_k$ for $k = 1, 2, \ldots$, $P(x) \equiv -xQ_i(x)$ and $g(t) \equiv e_{i0}(t)$. It is seen from (2.2.1) that $Q_i(0) = 1$. Moreover, by (2.2.4), $x_{1,i} > x_1 = \lim_{i\to\infty} x_{1,i} \geq 0$. Consequently, a $\delta > 0$ exists such that $Q_i(x) > 0$ for $x_1 \leq x \leq x_1 + \delta$. If $x_1 > 0$, then $-x_1Q_i(x_1) < 0$, and the lemma follows at once from theorem A.3.1. If, however, $x_1 = 0$, then, by (3.3.9) – (3.3.11) and theorem 2.2.6, $x_1 = x_k$ for all k, and theorem A.3.2 applies. □

The above lemma and theorem 5.1.3 manifestly lead to the next result.

THEOREM 5.2.2. *Let the natural birth-death process* $\{X_i(t)\}$, *with* $i > 0$, *be transient or null recurrent, i.e.,* $\Sigma\pi_n = \infty$. *Then* $\{X_i(t)\}$ *is strictly stochastically increasing on the interval* (τ_i, ∞), *where* τ_i *is the unique solution of the equation*

$$\int_0^\infty \exp(-xt)Q_i(x)xd\psi(x) = 0 .$$

5.3 The positive recurrent process

We have seen in section 3.3 that the process $\{X_i(t)\}$ is positive recurrent
iff $\Sigma \pi_n < \infty$.

LEMMA 5.3.1. *Let* $i > 0$ *and* $\Sigma \pi_n < \infty$. *Then* $e_{i0}(t) < 0$ *for* t *sufficiently large*
iff $Q_i(x_2) > 0$.

PROOF. As in lemma 5.2.1 the results of appendix 3 apply with $F \equiv \psi$, $u_k \equiv x_k$,
$k = 1, 2, \ldots$, $P(x) \equiv -xQ_i(x)$ and $g(t) \equiv e_{i0}(t)$.
From (3.3.11) one has $x_2 > x_1 = 0$. Thus if $Q_i(x_2) \neq 0$, it follows from
theorem A.3.1 with $\hat{u} = x_2$ that $\text{sign}(e_{i0}(t)) = -\text{sign}(Q_i(x_2))$ for t sufficiently
large.
If $Q_i(x_2) = 0$, i.e., $x_2 = x_{1,i}$, and $x_3 > x_2$, then, by lemma 2.2.3,
$x_{1,i} = x_2 < x_3 < x_{2,i}$, whence $Q_i(x_3) < 0$. Consequently, by theorem A.3.1,
$e_{i0}(t) > 0$ for t sufficiently large.
Finally, if $Q_i(x_2) = 0$ and $x_3 = x_2$, then, by theorem 2.2.6, $x_2 = x_k$ for all
$k > 2$ and, since $x_2 = x_{1,i}$, $Q_i(x_2+\delta) < 0$ for $\delta > 0$ sufficiently small. By
theorem A.3.2 it follows that $e_{i0}(t) > 0$ for t sufficiently large.   □

Theorem 5.1.3 and the preceding lemma imply the next important theorem.

THEOREM 5.3.2. *Let the natural birth-death process* $\{X_i(t)\}$, *with* $i > 0$, *be*
positive recurrent, i.e., $\Sigma \pi_n < \infty$. *Then* $\{X_i(t)\}$ *is strictly stochastically*
increasing in the long run iff $Q_i(x_2) > 0$. *Moreover, if* $Q_i(x_2) > 0$, *then*
$\{X_i(t)\}$ *is strictly stochastically increasing precisely on the interval* (τ_i, ∞),
where τ_i *is the unique solution of the equation*

$$\int_0^{\infty} \exp(-xt)Q_i(x)xd\psi(x) = 0 .$$

5.4 The case of an initial state distribution with finite support

If the initial distribution vector of a natural birth-death process is supported
by more than one point, the problem to determine whether the process is strictly
stochastically monotone in the long run becomes much less tractable. The main
reason for this is that, although a relation like (5.1.2) is valid, it no longer
holds in general that $S^-(\underline{e}(0)) \leq 1$, so that the sign of $e_0(t)$ for large values
of t does not determine any longer whether the process is strictly stochastically
monotone in the long run or not. But even when this sign would be conclusive, there

is still the problem of establishing its value, which is by no means easy if the initial distribution is supported by infinitely many points.

However, if the support of the initial distribution vector consists of finitely many points some conclusions can be drawn. They are expressed in the next two theorems.

THEOREM 5.4.1. *Consider a natural birth-death process* $\{X^*(t)\}$ *with* $\mu_0^* > 0$ *and initial distribution vector* $q^* = (q_{-1}^*, q_0^*, q_1^*, \ldots)^T$, *where* $q_{-1}^* < 1$ *and* $q_k^* = 0$ *for* k > n. *This process is not stochastically decreasing on any interval.*

PROOF. From lemma 4.2.1 one has

$$(5.4.1) \qquad e_i^*(t) = -\sum_{j=0}^{n} q_j^* e_{ij}(t) \ .$$

Considering lemma 4.3.1 *(ii)* it follows that for fixed t > 0 and i sufficiently large

$$e_i^*(t) < 0 \ .$$

On the other hand (4.3.2) implies that for t > 0

$$e_0^*(t) > 0$$

The theorem now follows from the results of section 4.2.   □

THEOREM 5.4.2. *Consider a natural birth-death process with* $\mu_0 = 0$ *and initial distribution vector* $q = (q_0, q_1, \ldots)^T$ *where* $q_k = 0$ *for* k > n.

(i)   *If the process is transient or null recurrent, i.e.,* $\Sigma \pi_n = \infty$, *then it is strictly stochastically increasing in the long run.*

(ii)   *If the process is positive recurrent, i.e.,* $\Sigma \pi_n < \infty$, *then it is not stochastically increasing on any interval if* $\Sigma q_i Q_i(x_2) < 0$.

PROOF. *(i):* From (4.1.10) it is seen that

$$(5.4.2) \qquad e(t) = q^T E(t) \ .$$

By lemma 5.1.2 the ith row of the matrix E(t) consists of negative elements iff the first element $e_{i0}(t)$ is negative, and by lemma 5.2.1 this occurs for every i provided t is sufficiently large. Since $q_i = 0$ for i > n, we conclude that $e(t)$ consists of negative elements for t sufficiently large, whence the result follows by theorem 4.1.3.

     (ii): Lemma 5.1.5 and (5.4.2) show that

$$e_0(t) = - \int_0^\infty \exp(-xt) \sum_{i=0}^{n} q_i Q_i(x) x \, d\psi(x) \ .$$

Since $\Sigma \pi_n < \infty$, we have $x_2 > x_1 = 0$, so that an appeal to theorem A.3.1 yields

$$e_0(t) > 0 \quad \text{for t sufficiently large}$$

given $\sum_{i=0}^{n} q_i Q_i(x_2) < 0$. On the other hand, lemma 4.3.1 *(i)* and the fact that $q_i = 0$ for $i > n$ are readily seen to imply that for fixed $t > 0$ and j sufficiently large

$$e_j(t) < 0 .$$

The theorem is now implied by the results of section 4.1.   ☐

6 THE M/M/s QUEUE LENGTH PROCESS

6.1 Introduction

In this chapter we consider the birth-death process $\{X(t): 0 \leq t < \infty\}$ with parameters

$$(6.1.1) \quad \begin{aligned} \lambda_n &= \lambda & n &= 0, 1, \ldots\ldots \\ \mu_n &= n\mu & n &= 0, 1, \ldots, s-1 \\ &= s\mu & n &= s, s+1, \ldots, \end{aligned}$$

which is the queue length process in an M/M/s queueing system.

It is readily seen that the potential coefficients of $\{X(t)\}$ are given by

$$(6.1.2) \quad \begin{aligned} \pi_n &= \rho^n s^n / n! & n &= 0, 1, \ldots, s-1 \\ &= \rho^n s^s / s! & n &= s, s+1, \ldots, \end{aligned}$$

where the *traffic intensity* ρ is defined by

$$(6.1.3) \quad \rho = \lambda / s\mu .$$

We note that

$$\sum_{n=0}^{\infty} (1/\lambda_n \pi_n) \sum_{i=0}^{n} \pi_i \geq \sum_{n=0}^{\infty} 1/\lambda_n = \infty$$

and

$$\sum_{n=0}^{\infty} (1/\lambda_n \pi_n) \sum_{i=n+1}^{\infty} \pi_i \geq \sum_{n=0}^{\infty} 1/\mu_{n+1} = \infty.$$

So the conditions C and D are satisfied, whence $\{X(t)\}$ is a natural birth-death process.

It further appears that

$$\Sigma \pi_n < \infty \qquad \text{iff } \rho < 1$$

and

$$\Sigma (1/\lambda_n \pi_n) < \infty \quad \text{iff } \rho > 1 .$$

Thus, by the results of section 3.3, the process $\{X(t)\}$ is

	transient	iff $\rho > 1$
(6.1.4)	null recurrent	iff $\rho = 1$
	positive recurrent	iff $\rho < 1$.

From (3.3.5) and (6.1.2) we obtain the well-known steady-state results for the positive recurrent case, viz.,

$$(6.1.5) \qquad \lim_{t\to\infty} p_{ij}(t) \equiv p_j = p_0 \rho^j s^j/j! \qquad\qquad j = 0, 1, \ldots, s-1$$

$$= p_0 \rho^j s^s/s! \qquad\qquad j = s, s+1, \ldots.$$

and

$$(6.1.6) \qquad p_0 = 1/\{ \sum_{j=0}^{s-1}((\rho s)^j/j!) + (\rho s)^s/(1-\rho)s!\}.$$

We have seen in section 2.2 that the transition probability function $p_{ij}(t)$, $0 \le t < \infty$, is represented by

$$(6.1.7) \qquad p_{ij}(t) = \pi_j \int_0^\infty \exp(-xt)Q_i(x)Q_j(x)d\psi(x) \ .$$

KARLIN and McGREGOR (1958[a]) have shown that

$$(6.1.8) \qquad Q_n(x) = c_n(x/\mu, \lambda/\mu) \qquad\qquad n = 0, 1, \ldots, s \ ,$$

where $\{c_n(x,a)\}_n$ are the *Poisson-Charlier polynomials*, defined by

$$c_0(x,a) = 1$$
$$(6.1.9) \qquad -xc_0(x,a) = -ac_0(x,a) + ac_1(x,a)$$
$$-xc_n(x,a) = nc_{n-1}(x,a) - (n+a)c_n(x,a) + ac_{n+1}(x,a) \qquad n > 0.$$

ERDELYI (1953) states that

$$(6.1.10) \qquad c_n(x,a) = \sum_{r=0}^n (-1)^r \binom{n}{r}\binom{x}{r}r!/a^r \ .$$

Consequently,

$$(6.1.11) \qquad Q_n(x) = \sum_{r=0}^n (-1)^r \binom{n}{r}\binom{\rho s x/\lambda}{r}r!/(\rho s)^r \qquad n = 0, 1, \ldots, s \ .$$

The polynomials $\{Q_n(x)\}$ for $n > s$ are given by

$$(6.1.12) \qquad Q_{s+n}(x) = (1/\sqrt\rho)^n\{Q_s(x)U_n(\alpha(x)\sqrt\rho) - Q_{s-1}(x)U_{n-1}(\alpha(x)\sqrt\rho)/\sqrt\rho\}$$

$$n = 1, 2, \ldots,$$

where

(6.1.13) $\alpha(x) = \tfrac{1}{2}(1 - x/\lambda + 1/\rho)$

and $\{U_n(x)\}_n$ are the *Chebysev polynomials of the second kind* defined by

(6.1.14)
$$U_{-1}(x) = 0, \quad U_0(x) = 1$$
$$xU_n(x) = \tfrac{1}{2}U_{n-1}(x) + \tfrac{1}{2}U_{n+1}(x) \qquad\qquad n = 0, 1, \ldots\ldots .$$

According to ERDÉLYI (1953) we have

(6.1.15) $U_n(\cos\theta) = \sin(n+1)\theta/\sin\theta$.

KARLIN and McGREGOR (1958[a]) have deduced certain general features of the spectral function of $\{X(t)\}$. However, their analysis is not so thorough as desirable and lacks clarity. The purpose of the next section is therefore to improve and clarify KARLIN and McGREGOR's analysis and results. Our starting-point will be the formula for the Stieltjes transform of the spectral function of the M/M/s queue length process as derived by KARLIN and McGREGOR (1958[a]).

6.2   The spectral function

In view of (6.1.3) and (6.1.13) the recurrence formulas (2.2.1) for the poly-nomials belonging to the M/M/s queue length process $\{X(t)\}$ can be written as

(6.2.1)
$$Q_0(x) = 1 \ , \ Q_1(x) = 1 - x/\lambda$$
$$Q_{n+1}(x) = (1 - x/\lambda + n/\rho s)Q_n(x) - nQ_{n-1}(x)/\rho s \qquad 0 < n < s$$
$$Q_{n+1}(x) = 2\alpha(x)Q_n(x) - Q_{n-1}(x)/\rho \qquad\qquad n \geq s.$$

The polynomials defined by

(6.2.2)
$$Q_0^{(0)}(x) = 0 \ , \ Q_1^{(0)}(x) = -1/\lambda$$
$$Q_{n+1}^{(0)}(x) = (1 - x/\lambda + n/\rho s)Q_n^{(0)}(x) - nQ_{n-1}^{(0)}(x)/\rho s \qquad 0 < n < s.$$
$$Q_{n+1}^{(0)}(x) = 2\alpha(x)Q_n^{(0)}(x) - Q_{n-1}^{(0)}(x)/\rho \qquad\qquad n \geq s,$$

are called the associated polynomials of the system $\{Q_n(x)\}_n$.
According to KARLIN and McGREGOR (1958[a]) the Stieltjes transform of the spectral function ψ of $\{X(t)\}$ is given by

(6.2.3) $B(z) \equiv \int\limits_0^\infty d\psi(x)/(x-z) = -N(z)/D(z) \ , \ 0 < \arg z < 2\pi, \ |z| > 0,$

where

(6.2.4) $\qquad N(z) = C(z)Q_{s-1}^{(0)}(z) - Q_s^{(0)}(z)$

(6.2.5) $\qquad D(z) = C(z)Q_{s-1}(z) - Q_s(z)$

and

(6.2.6) $\qquad C(z) = P.V.(\alpha(z) - \sqrt{\alpha^2(z)-1/\rho})$

(P.V. denotes principle value). We note that $C(z)$ is holomorphic in the z-plane cut along the interval $[b_1,b_2]$, where

(6.2.7) $\qquad b_1 = \lambda(1 - 1/\sqrt{\rho})^2 \;, \; b_2 = \lambda(1 + 1/\sqrt{\rho})^2$

are the branch points of the double-valued function

$$\sqrt{\alpha^2(z) - 1/\rho} \; .$$

Considering that Q_{s-1}, Q_s, $Q_{s-1}^{(0)}$ and $Q_s^{(0)}$ are polynomials, we see from (6.2.4) - (6.2.6) that the function

(6.2.8) $\qquad H(z) = -N(z)/D(z)$

is holomorphic in the z-plane cut along the interval $[b_1,b_2]$, with the exception of the zeros, if any, of $D(z)$.

It is easily seen by induction that for all z and $n = 1, 2, \ldots, s$,

(6.2.9) $\qquad Q_n(z)Q_{n-1}^{(0)}(z) - Q_n^{(0)}(z)Q_{n-1}(z) = (n-1)!/\lambda(\rho s)^{n-1} \;,$

which implies

(6.2.10) $\qquad Q_s(z)Q_{s-1}^{(0)}(z) \neq Q_s^{(0)}(z)Q_{s-1}(z)$

for all z. Therefore, a zero of $D(z)$ is definitely not a zero of $N(z)$, whence each zero of $D(z)$ is a non-removable singularity of $H(z)$.

We shall now look into the position of the zeros of $D(z)$. We have stated in sector 2.2 that $B(z)$ is holomorphic in the entire z-plane with the non-negative real axis removed. Hence zeros of $D(z)$, which are non-removable singularities of $H(z)$, the analytic continuation of $B(z)$, are on the non-negative real axis. More precise statements are possible, however. First suppose

$$b_1 < x < b_2 \; .$$

Then, clearly, $\text{Im } D(x) \neq 0$ unless $Q_{s-1}(x) = 0$. Consequently, if $D(x) = 0$, then $Q_{s-1}(x) = 0$, whence

$$0 = D(x) = C(x)Q_{s-1}(x) - Q_s(x) = -Q_s(x) \ .$$

But $Q_{s-1}(x) = Q_s(x) = 0$ contradicts (6.2.10). Therefore, zeros of $D(z)$ cannot occur in the real interval (b_1, b_2).

Zeros of $D(z)$ cannot occur in the real interval $[b_2, \infty)$ either, as can be seen by the following argument. Define

$$(6.2.11) \qquad R_n(x) = Q_n(x)/Q_{n-1}(x) \qquad\qquad n = 1, 2, \ldots \ .$$

In view of (6.2.1) we have

$$R_1(x) = 1 - x/\lambda$$

$$(6.2.12) \qquad R_{n+1}(x) = 1 - x/\lambda + n/\rho s - n/\rho s R_n(x) \qquad\qquad 0 < n < s$$

$$R_{n+1}(x) = 2\alpha(x) - 1/\rho R_n(x) \qquad\qquad n \geq s \ .$$

We shall show that for any $x \geq b_2$ and all $n = 1, 2, \ldots\ldots$, s the following proposition holds:

$$(6.2.13) \qquad -\infty < R_n(x) < C(x) < 0 \ .$$

Namely, if

$$x \geq b_2 = \lambda(1 + 1/\sqrt{\rho})^2$$

then

$$\alpha(x) = \tfrac{1}{2}(1 - x/\lambda + 1/\rho) \leq -1/\sqrt{\rho} \ .$$

Consequently,

$$\alpha(x) = -\sqrt{\alpha^2(x)} < -\sqrt{\alpha^2(x) - 1/\rho} < 0 \ ,$$

whence

$$2\alpha(x) < \alpha(x) - \sqrt{\alpha^2(x) - 1/\rho} = C(x) < \alpha(x) < 0 \ .$$

Since

$$R_1(x) = 1 - x/\lambda < 2\alpha(x) \ ,$$

it follows that proposition (6.2.13) is valid for $n = 1$. Now suppose it is valid for $n = 1, \ldots, m < s$, then

$$(6.2.14) \qquad 1/C(x) < 1/R_m(x) < 0.$$

Moreover, one readily sees that

(6.2.15) $\qquad 2\alpha(x) = C(x) + 1/\rho C(x)$,

so that

(6.2.16) $\qquad 1 - x/\lambda = C(x) - 1/\rho + 1/\rho C(x)$.

Combining this result with (6.2.14) yields

$$-\infty < R_{m+1}(x) = 1 - x/\lambda + m/\rho s - m/\rho s R_m(x) <$$
$$< C(x) - 1/\rho + 1/\rho C(x) + m/\rho s - m/\rho s C(x) =$$
$$= C(x) - (1 - m/s)(1 - 1/C(x))/\rho < C(x) < 0 \ .$$

We conclude that proposition (6.2.13) holds for all $n \leq s$ and $x \geq b_2$, so that in particular

(6.2.17) $\qquad C(x) - R_s(x) > 0 \ \text{if} \ x \geq b_2$.

From (6.2.5) and (6.2.11) we have

(6.2.18) $\qquad D(x) = Q_{s-1}(x)(C(x) - R_s(x))$.

It is evident from (6.2.5) and (6.2.10) that $D(x)$ and $Q_{s-1}(x)$ do not have common zeros. Thus, by (6.2.18), zeros of $D(x)$ are zeros of $C(x) - R_s(x)$ and these do not occur for $x \geq b_2$ as we have just shown.
Summarizing the above results we obtain the following.

LEMMA 6.2.1. *Zeros of* $D(z)$ *can occur only in the real interval* $[0,b_1]$.

It will be clear that the (finitely many) zeros of $D(z)$ are poles of $1/D(z)$, if outside the interval $[b_1,b_2]$. Considering lemma 6.2.1 and our previous result that a zero of $D(z)$ is a non-removable singularity of $H(z) = -N(z)/D(z)$, it follows that the zeros $\neq b_1$ of $D(z)$ are precisely the poles of $H(z)$. Moreover, the latter function has no other singularities besides these poles in the z-plane cut along the interval $[b_1,b_2]$. At this point we can state the next lemma, by which (following the procedure suggested in section 2.2) a first step has been taken towards a solution of the problem of finding the spectral function ψ.

LEMMA 6.2.2. $H(z) = -N(z)/D(z)$ *is the analytic continuation of* $B(z)$ *in the entire z-plane cut along the interval* $[b_1,b_2]$ *with the exception of the zeros, if any, of* $D(z)$, *which lie in the real interval* $[0,b_1]$.

We will now study the behaviour of $H(z)$ in a neighbourhood of a zero of $D(z)$. From (2.2.1) we easily obtain with induction

$$\lambda_n \pi_n (Q_n(x) Q'_{n-1}(x) - Q'_n(x) Q_{n-1}(x)) = \sum_{i=0}^{n-1} \pi_i Q_i^2(x),$$

so that

$$(6.2.19) \qquad Q'_n(x) Q_{n-1}(x) - Q_n(x) Q'_{n-1}(x) < 0$$

for all $n = 1, 2, \ldots$ and real x. It follows that

$$(6.2.20) \qquad -\infty < R'_s(x) = (d/dx)(Q_s(x)/Q_{s-1}(x)) < 0$$

for all real x such that $Q_{s-1}(x) \neq 0$. Furthermore, it is easy to verify from (6.2.6) that for any real $x < b_1$

$$(6.2.21) \qquad \infty > C'(x) = C(x)/\sqrt{(b_1-x)(b_2-x)} > 0 .$$

Finally, we see

$$(6.2.22) \qquad C'(x) \to \infty \text{ as } x \uparrow b_1 .$$

Now suppose $\hat{x} \neq b_1$ is a zero of $D(z)$. Then, by lemma 6.2.1, $0 \leq \hat{x} < b_1$. Since $D(z)$ and $Q_{s-1}(z)$ do not have common zeros, it is implied by (6.2.18) that

$$C(\hat{x}) - R_s(\hat{x}) = 0.$$

Further, we have by (6.2.20) and (6.2.21)

$$0 < C'(\hat{x}) - R'_s(\hat{x}) < \infty .$$

Consequently,

$$D'(\hat{x}) = (d/dz)\{Q_{s-1}(z)(C(z) - R_s(z))\}_{z=\hat{x}} =$$
$$= Q_{s-1}(\hat{x})(C'(\hat{x}) - R'_s(\hat{x})) \neq 0 .$$

It follows that

$$D(z) = (z - \hat{x})F(z) ,$$

with $F(z)$ holomorphic in a neighbourhood of $\hat{x}$ and $F(\hat{x}) \neq 0$. Our conclusion is expressed in the next lemma.

LEMMA 6.2.3. *If* $\hat{x} \neq b_1$ *and* $D(\hat{x}) = 0$, *then* $H(z) = -N(z)/D(z)$ *has a simple pole in* $\hat{x}$.

Finally, suppose $D(b_1) = 0$. An argument similar to that above then yields

$$C(b_1) - R_s(b_1) = 0 \ .$$

Hence, considering that b_1 is a second order branch point of $C(z)$, we can write

(6.2.23) $C(z) - R_s(z) = \sqrt{(b_1-z)}F_1(z) + (b_1-z)F_2(z) \ ,$

with F_1 and F_2 holomorphic in a neighbourhood of b_1. From (6.2.10) and (6.2.22) we get

$$C'(x) - R_s'(x) \to \infty \text{ as } x \uparrow b_1 \ , \ x \text{ real.}$$

Meanwhile (6.2.23) implies, for z not in the real interval $[b_1,\infty)$

$$C'(z) - R_s'(z) = -(1/2\sqrt{(b_1-z)})F_1(z) + \sqrt{(b_1-z)}F_1'(z) + F_3(z) \ ,$$

with F_3 holomorphic in a neighbourhood of b_1. It follows that $F_1(b_1) \neq 0$. If we write

(6.2.24) $D(z) = \sqrt{(b_1-z)}F_4(z) \ ,$

then, by (6.2.18) and (6.2.23),

$$F_4(z) = Q_{s-1}(z)(F_1(z) + \sqrt{(b_1-z)}F_2(z)) \ ,$$

so that

(6.2.25) $F_4(b_1) = Q_{s-1}(b_1)F_1(b_1) \neq 0 \ .$

Considering that $N(b_1) \neq 0$ if $D(b_1) = 0$, the next lemma is now obvious.

LEMMA 6.2.4. *If* $D(b_1) = 0$, *then* $H(z) = -N(z)/D(z) = O(1/\sqrt{b_1-z}),(z \to b_1)$.

The lemmas 6.2.1 - 4 together with theorem 2.2.8 enable us to describe the behaviour of the spectral function ψ. We discern four cases.

I   Theorem 2.2.8 and lemma 6.2.2 imply that for x outside the interval $[b_1,b_2]$ and unequal to one of the zeros of $D(z)$,

$$\psi'(x) = 0 \ .$$

II   Theorem 2.2.8 and lemma 6.2.3 imply that if x is one of the finitely many zeros of $D(z)$ in the interval $[0,b_1)$, then ψ has a jump in x of magnitude

$$\Delta\psi(x) = \text{Res}_x N(z)/D(z) \ ,$$

whence by previous results

$$\Delta\psi(x) = N(x) \lim_{z \to x} (z-x)/D(z) > 0 .$$

III $D(z)$ has no zeros in the interval $(b_1,b_2]$ by lemma 6.2.1, therefore

$$\text{Im } B(\xi+i\eta) = \text{Im } H(\xi+i\eta) = -\text{Im } N(\xi+i\eta)/D(\xi+i\eta)$$

converges uniformly to some continuous function $f(\xi)$ as $\eta \downarrow 0$ on any closed interval which is contained in the interval $(b_1,b_2]$. Consequently, the inversion formula (2.2.7) implies

(6.2.26) $\psi'(x) = f(x)/\pi$

if x is in such a closed interval. To determine f we define

(6.2.27) $\Gamma(z) = P.V.(\alpha(z) + \sqrt{\alpha^2(z)-1/\rho})$

in the z-plane cut along the interval $[b_1,b_2]$. It is easy to verify that

(6.2.28) $C(z)\Gamma(z) = 1/\rho$

and

(6.2.29) $C(z) + \Gamma(z) = 2\alpha(z) .$

Moreover, in a manner analogous to that of the proof of our earlier statement

$$D(z) \equiv C(z)Q_{s-1}(z) - Q_s(z) \neq 0$$

for z in the interval (b_1,b_2), one can show that

(6.2.30) $\Gamma(z)Q_{s-1}(z) - Q_s(z) \neq 0$

for z in the same interval. Since $\Gamma(b_2) = C(b_2)$ and $D(b_2) \neq 0$ by lemma 6.2.1, the inequality (6.2.30) is valid even for all z in the interval $(b_1,b_2]$. Hence, for $b_1 < \xi \le b_2$,

$$f(\xi) = \lim_{\eta \downarrow 0} \text{Im } B(\xi+i\eta) = -\text{Im } N(\xi)/D(\xi) =$$

$$= -\text{Im } \left\{ \frac{C(\xi)Q_{s-1}^{(0)}(\xi) - Q_s^{(0)}(\xi)}{C(\xi)Q_{s-1}(\xi) - Q_s(\xi)} \times \frac{\Gamma(\xi)Q_{s-1}(\xi) - Q_s(\xi)}{\Gamma(\xi)Q_{s-1}(\xi) - Q_s(\xi)} \right\} =$$

$$= \frac{\sqrt{1/\rho - \alpha^2(\xi)}\{Q_{s-1}^{(0)}(\xi)Q_s(\xi) - Q_s^{(0)}(\xi)Q_{s-1}(\xi)\}}{\{Q_{s-1}(\xi)/\rho - 2\alpha(\xi)Q_s(\xi)\}Q_{s-1}(\xi) + Q_s^2(\xi)} .$$

Considering (6.2.1) and (6.2.9) this reduces to

(6.2.31) $f(\xi) = s! \sqrt{\rho - \rho^2\alpha^2(\xi)}/\lambda(\rho s)^s\{Q_s^2(\xi) - Q_{s-1}(\xi)Q_{s+1}(\xi)\}$.

Thus, by (6.2.26) and (6.2.31), one has for $b_1 < x \le b_2$

(6.2.32) $\psi'(x) = s! \sqrt{\rho - \rho^2\alpha^2(x)}/\lambda\pi(\rho s)^s\{Q_s^2(x) - Q_{s-1}(x)Q_{s+1}(x)\}$.

IV   By lemma 6.2.2 there remains only one point to consider, namely b_1.
If no zero of $D(z)$ coincides with the branch point b_1, then (6.2.32) is easily
seen to hold for $x=b_1$ as well, in which case

(6.2.33) $\psi'(b_1) = 0$.

Next suppose $D(b_1) = 0$. Choose $\delta > 0$ such that no zero of $D(z)$ occurs in the
interval $[b_1-\delta,b_1)$. Then ψ is constant on this interval as we have seen. The
inversion formula (2.2.7) now yields

(6.2.34) $\frac{1}{2}(\psi(b_1+0) - \psi(b_1)) = \frac{1}{2}(\psi(b_1+0) + \psi(b_1)) - \psi(b_1-\delta) =$

$$= \lim_{\eta \downarrow 0}(1/\pi) \int_{b_1-\delta}^{b_1} \text{Im } B(\xi+i\eta)d\xi = \lim_{\eta \downarrow 0}(1/\pi) \int_{b_1-\delta}^{b_1} \text{Im } H(\xi+i\eta)d\xi \quad .$$

Using the relation (2.2.10) we obtain the following equation (cf. figure 6.2.1).

$$(1/\pi) \int_{b_1-\delta}^{b_1} \text{Im } H(\xi+i\eta)d\xi = (1/2\pi i) \int_{b_1-\delta}^{b_1} (H(\xi+i\eta)-H(\xi-i\eta))d\xi =$$

$$= (1/2\pi i) \int_{b_1-\delta+i\eta}^{b_1+i\eta} H(\xi)d\xi + (1/2\pi i) \int_{b_1-i\eta}^{b_1-\delta-i\eta} H(\xi)d\xi =$$

(6.2.35) $= -(1/2\pi i) \int_{b_1-\delta-i\eta}^{b_1-\delta+i\eta} H(\xi)d\xi - (1/2\pi i) \int_L H(\xi)d\xi \quad ,$

where L is the half-circle $b_1 + \eta\exp(i\phi)$ $(\pi/2 \le \phi \le 3\pi/2)$ oriented from
$b_1+i\eta$ to $b_1-i\eta$.

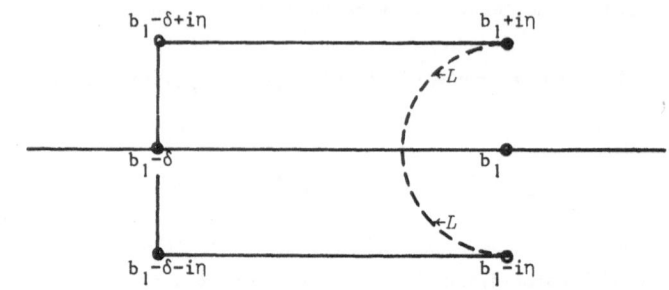

figure 6.2.1

The first integral in (6.2.35) tends to zero as $\eta \downarrow 0$, since H(z) is analytic on the path of integration. As a consequence of lemma 6.2.4, $H(z) = 0\,(1/\sqrt{\eta})$ on L. Thus, considering that the length of L equals $\pi\eta$, also the second integral in (6.2.35) tends to zero as $\eta \downarrow 0$. Substitution of these results in (6.2.34) yields

$$\psi(b_1+0) - \psi(b_1) = 0 .$$

We conclude that (6.2.33) holds, whether b_1 is a zero of D(z) or not.

Collecting our results we have the following theorem.

THEOREM 6.2.5. *For the spectral function ψ of the M/M/s queue length process the following holds:*

If $\lambda(1 - 1/\sqrt{\rho})^2 \le x \le \lambda(1 + 1/\sqrt{\rho})^2$, *then*

$$\psi'(x) = s!\sqrt{\rho - \rho^2\alpha^2(x)}/\lambda\pi(\rho s)^s\{Q_s^2(x) - Q_{s-1}(x)Q_{s+1}(x)\} .$$

If $x \ne \lambda(1 - 1/\sqrt{\rho})^2$ *is a zero of D(z), then ψ has a jump in x of magnitude*

$$\Delta\psi(x) = N(x) \lim_{z \to x} (z-x)/D(z) > 0 .$$

For any other value of x, $\psi'(x) = 0$. N(.), D(.) *and* $\alpha(.)$ *are given by (6.2.4), (6.2.5) and (6.1.13), respectively.*

The remaining part of this section will be concerned with the computation of the number of jumps of the spectral function ψ. We have seen that these jumps occur precisely at the zeros $\ne b_1$ of $D(x) = Q_{s-1}(x)(C(x) - R_s(x))$.
Since D(x) and $Q_{s-1}(x)$ do not have common zeros, the jumps occur precisely at the

zeros $\neq b_1$ of $C(x) - R_s(x)$. Further, it has been shown that the zeros of
$C(x) - R_s(x)$ occur in the interval $[0,b_1]$
In what follows we shall need some knowledge of the shape of the functions $C(.)$
and $R_s(.)$. It is seen by (6.2.6) and (6.2.21) that $C(x)$ is increasing on the
interval $[0,b_1]$, with

$$(6.2.36) \qquad C(0) = \tfrac{1}{2}(1 + 1/\rho - |1 - 1/\rho|) = 1 \qquad\qquad \text{if} \quad 0 < \rho \leq 1$$

$$= 1/\rho \qquad\qquad \text{if} \qquad \rho \geq 1$$

and

$$(6.2.37) \qquad C(b_1) = C(\lambda(1 - 1/\sqrt{\rho})^2) = 1/\sqrt{\rho} \ .$$

As regards the function $R_s(x) = Q_s(x)/Q_{s-1}(x)$ we note the following.
It was stated in section 2.2 that the n zeros of the polynomial $Q_n(x)$ are
distinct and on the positive real axis. They are denoted by $x_{i,n}$,
$i = 1, 2, \ldots, n$, and numbered such that $x_{i,n} < x_{i+1,n}$. In view of this and
(6.2.20) it is clear that on the interval $[0,x_{1,s-1})$ the function $R_s(x)$
decreases from $R_s(0) = 1$ to $-\infty$, and on the intervals $(x_{i,s-1}, x_{i+1,s-1})$, $i = 1,\ldots,$
$s-1$, from $+\infty$ to $-\infty$ (we let $x_{s,s-1} \equiv \infty$). Finally, it is also easy to see that
on the interval $(-\infty,0]$ the function $R_s(x)$ decreases from $+\infty$ to 1.
The number of zeros of $C(x) - R_s(x)$ obviously depends on the value of the traffic
intensity ρ, since in fact, $C(x) = C(x,\rho)$ and $R_s(x) = R_s(x,\rho)$. From now on we
shall indicate every dependence on ρ.
Our first main step in the computation of the exact number of zeros of
$C(x,\rho) - R_s(x,\rho)$ for a fixed $\rho > 0$, will be the determination of the number of
zeros of this function on the curve

$$x = b_1(\rho) = \lambda(1 - 1/\sqrt{\rho})^2 \qquad\qquad\qquad \rho > 0,$$

i.e., the number of positive real zeros of

$$(6.2.38) \qquad 1/\sqrt{\rho} - R_s(b_1(\rho),\rho) \ .$$

An upperbound for this number is easily established.

LEMMA 6.2.6. *The function* $1/\sqrt{\rho} - R_s(b_1(\rho),\rho)$ *has at most s positive real zeros.*

PROOF. It is easily seen from (6.2.1) and (6.2.7) that for $n \leq s$ the function
$(\sqrt{\rho})^n Q_n(b_1(\rho),\rho)$ is a polynomial of degree n in $1/\sqrt{\rho}$. Hence the equation

$$(\sqrt{\rho})^{s-1} Q_{s-1}(b_1(\rho),\rho) \quad - (\sqrt{\rho})^s Q_s(b_1(\rho),\rho) = 0$$

has at most s positive roots. The result follows immediately. $\qquad\qquad\qquad \square$

To obtain a lowerbound for the number of positive real zeros of $1/\sqrt{\rho} - R_s(b_1(\rho),\rho)$ we need the preparatory lemmas 6.2.7 and 6.2.8.

LEMMA 6.2.7. $R_s(x,\rho) < -1/\sqrt{\rho}$ for $x \geq b_1(\rho)$ and $\rho > 0$ sufficiently small.

PROOF. Let $0 < \rho < 1/16s^2$ and $x \geq b_1(\rho) = \lambda(1-1/\sqrt{\rho})^2$. If

$$R_n(x,\rho) < -1/\sqrt{\rho} ,$$

with $n < s$, then, by (6.2.12),

$$R_{n+1}(x,\rho) = 1 - x/\lambda + n/\rho s - n/\rho s R_n(x,\rho) <$$

$$< 1 - (1-1/\sqrt{\rho})^2 + n/\rho s + n/s\sqrt{\rho} =$$

$$= (2 + n/s)/\sqrt{\rho} - (1-n/s)/\rho <$$

$$< -1/\sqrt{\rho} + (4 - (1-n/s)/\sqrt{\rho})/\sqrt{\rho} \leq -1/\sqrt{\rho} ,$$

since for $n < s$,

$$(1 - n/s)/\sqrt{\rho} \geq 1/s\sqrt{\rho} \geq 4.$$

Now

$$R_1(x,\rho) = 1 - x/\lambda \leq 1 - (1-1/\sqrt{\rho})^2 =$$

$$= -1/\sqrt{\rho} - (1/\sqrt{\rho} - 3)/\sqrt{\rho} < -1/\sqrt{\rho} .$$

Thus by induction $R_s(x,\rho) < -1/\sqrt{\rho}$. $\qquad\qquad\qquad\qquad\qquad$ $\square$

LEMMA 6.2.8. $R_s(x,\rho) > 1/\sqrt{\rho}$ for $x \leq b_1(\rho)$ and $\rho > 1$.

PROOF. Let $\rho > 1$ and $x \leq b_1(\rho) = \lambda(1-1/\sqrt{\rho})^2$. If

$$R_n(x,\rho) > 1/\sqrt{\rho} ,$$

with $n < s$, then, by (6.2.12),

$$R_{n+1}(x,\rho) = 1 - x/\lambda + n/\rho s - n/\rho s R_n(x,\rho) >$$

$$> 1 - (1-1/\sqrt{\rho})^2 + n/\rho s - n/s\sqrt{\rho} =$$

$$= 1/\sqrt{\rho} + (1-n/s)(1-1/\sqrt{\rho})/\sqrt{\rho} > 1/\sqrt{\rho} .$$

Now

$$R_1(x,\rho) = 1 - x/\lambda \geq 1 - (1 - 1/\sqrt{\rho})^2 =$$

$$1/\sqrt{\rho} + (1 - 1/\sqrt{\rho})/\sqrt{\rho} > 1/\sqrt{\rho} \ .$$

Thus by induction $R_s(x,\rho) > 1/\sqrt{\rho}$.   □

Thirdly, we need the following. Considering the behaviour of $R_s(x,\rho)$
as a function of x (which has been described earlier), the function

(6.2.39) $1/\sqrt{\rho} - R_s(x,\rho)$

has exactly s distinct real zeros

(6.2.40) $z_1(\rho) < z_2(\rho) < \ldots < z_s(\rho)$,

which have the following, easily verifiable properties.

$$z_1(\rho) < 0 \text{ iff } 0 < \rho < 1$$

(6.2.41) $z_1(\rho) = 0 \text{ iff } \quad \rho = 1$

$$z_1(\rho) > 0 \text{ iff } \quad \rho > 1 \ ,$$

and, for $i = 1, 2, \ldots, s$,

(6.2.42) $x_{i-1,s-1}(\rho) < z_i(\rho) < x_{i,s-1}(\rho)$,

where $x_{0,s-1}(\rho) \equiv -\infty$ and $x_{s,s-1}(\rho) \equiv \infty$. Finally, for $\rho > 0$ and $i = 1, 2, \ldots, s$,

(6.2.43) $z_i(\rho)$ is differentiable,

which follows from a well-known theorem on implicit functions and the fact
that

$$-\infty < (\partial/\partial x)R_s(x,\rho) < 0$$

in a neighbourhood of $x = z_i(\rho)$, as a consequence of (6.2.20) and (6.2.42).

LEMMA 6.2.9. *The function* $1/\sqrt{\rho} - R_s(b_1(\rho),\rho)$ *has at least* s *positive real
zeros and these are in the interval* $(0,1]$.

PROOF. Lemma 6.2.7 implies

$$1/\sqrt{\rho} - R_s(x,\rho) > 2/\sqrt{\rho} > 0$$

for $x \geq b_1(\rho)$ and $\rho > 0$ sufficiently small. It follows that

$$z_s(\rho) < b_1(\rho)$$

for $\rho > 0$ sufficiently small. From lemma 6.2.8 we see that

$$z_1(\rho) > b_1(\rho)$$

for $\rho > 1$. Considering (6.2.40) and the fact that each $z_i(\rho)$ is continuous by (6.2.43), we conclude that the graph of $z_i(\rho)$, $i = 1, 2, \ldots, s$, intersects the graph of $b_1(\rho)$ at least once on the interval $(0,1]$, but nowhere else. The lemma follows at once. □

We remark that $\rho = 1$ is a zero of the function (6.2.38) since $R_s(b_1(1),\rho) = R_s(0,\rho) = 1$. This observation together with the lemmas 6.2.6 and 6.2.9 leads to the next result.

LEMMA 6.2.10. *The function $1/\sqrt{\rho} - R_s(b_1(\rho),\rho)$ has exactly s distinct positive zeros ρ_i, $i = 1, 2, \ldots, s$. If they are numbered such that $\rho_{i+1} < \rho_i$, then also*

$$0 < \rho_s < \rho_{s-1} < \ldots < \rho_1 = 1 .$$

We return to the original problem of finding the number of jumps of $\psi = \psi(x,\rho)$, i.e., the number of zeros of $C(x,\rho) - R_s(x,\rho)$ in the interval $[0,b_1(\rho))$, for any fixed $\rho > 0$. For $\rho \geq \rho_1 = 1$ we have the following result.

LEMMA 6.2.11. *If $\rho \geq 1$, then $\psi(x,\rho)$ is continuous.*

PROOF. It is easily verified that $R_s(x,\rho) > 1/\sqrt{\rho}$ if $\rho \geq 1$ and $x < b_1(\rho)$, the argument being similar to the proof of lemma 6.2.8. Considering (6.2.37) and the fact that $C(x,\rho)$ is increasing with x on $[0,b_1(\rho)]$, we therefore have for $x < b_1(\rho)$ and $\rho \geq 1$

$$C(x,\rho) - R_s(x,\rho) < C(b_1(\rho),\rho) - R_s(x,\rho) =$$

$$= 1/\sqrt{\rho} - R_s(x,\rho) < 0 .$$

Thus $C(x,\rho) - R_s(x,\rho)$ has no zeros for $x < b_1(\rho)$ if $\rho \geq 1$, whence $\psi(x,\rho)$ has no jumps if $\rho \geq 1$. □

For ρ in the interval $(0,1)$ we discern s cases, viz., $0 < \rho < \rho_s$ and $\rho_{i+1} \leq \rho < \rho_i$, $i = 1, 2, \ldots, s-1$, which can be analysed simultaneously.

LEMMA 6.2.12. *If $\rho > 0$ and $\rho_{i+1} \leq \rho < \rho_i$, with $1 \leq i \leq s$ and $\rho_{s+1} \equiv 0$, then $\psi(x,\rho)$ has exactly i jumps.*

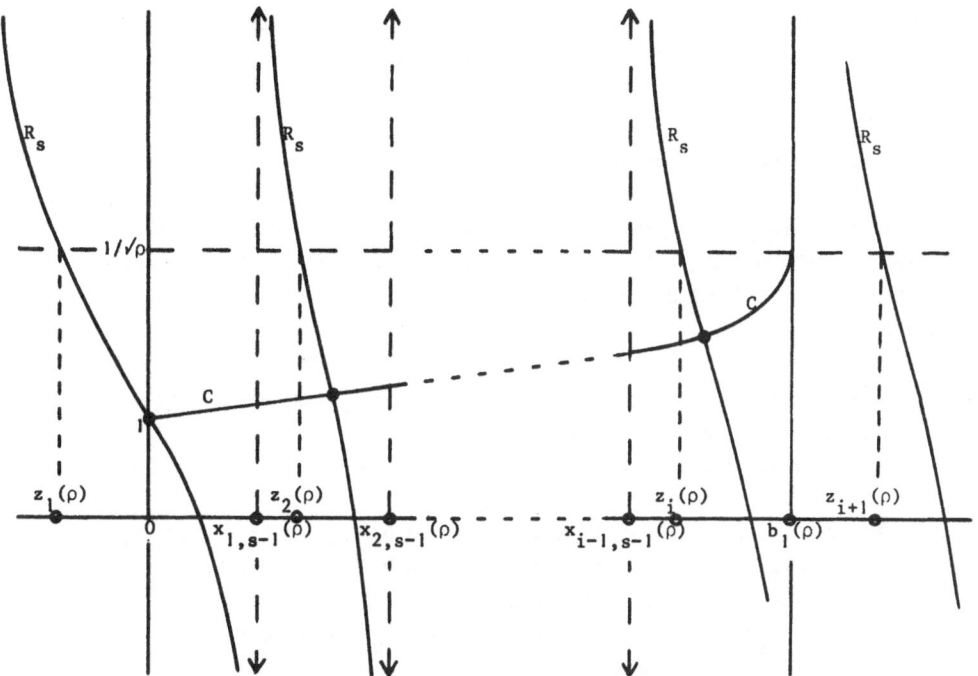

figure 6.2.2

PROOF (cf. figure 6.2.2). Let $\rho > 0$ and $\rho_{i+1} \leq \rho < \rho_i$, then

(6.2.44) $z_1(\rho) < 0 < x_{1,s-1}(\rho) < z_2(\rho) < \ldots < x_{i-1,s-1}(\rho) < z_i(\rho) < b_1(\rho)$;

additionally, if $i < s$,

(6.2.45) $b_1(\rho) \leq z_{i+1}(\rho)$

and

(6.2.46) $z_i(\rho) < x_{i,s-1}(\rho)$.

Namely, lemma 6.2.7 implies for $k = 1, 2, \ldots\ldots,$

$z_k(\sigma) < b_1(\sigma)$ for $\sigma > 0$ sufficiently small,

while as consequence of lemma 6.2.8

$z_k(\sigma) > b_1(\sigma)$ for $\sigma > 1$

(cf. the proof of lemma 6.2.9). Considering that the graphs of $z_k(\sigma)$ and $b_1(\sigma)$ intersect exactly once, viz., at ρ_k, it follows that the last inequalities of (6.2.44) and (6.2.45) are valid. For the other inequalities see (6.2.41) and (6.2.42).

In view of the behaviour of $C(x,\rho)$ and $R_s(x,\rho)$ as functions of x on the interval $[0,b_1(\rho)]$, as mentioned previously, it is now readily verified that the graph of $C(x,\rho)$ intersects the graph of $R_s(x,\rho)$ exactly once on each interval $[z_j(\rho),z_{j+1}(\rho)) \cap [0,b_1(\rho))$, j = 1, 2, ..., i. Thus $C(x,\rho) - R_s(x,\rho)$ has exactly i zeros in the interval $0 \le x < b_1(\rho)$ if $\rho > 0$ and $\rho_{i+1} \le \rho < \rho_i$, whence the lemma follows.   □

We remark that for $0 < \rho < 1$, the function $C(x,\rho) - R_s(x,\rho)$ has a zero at x = 0, so $\psi(x,\rho)$ has a jump at x = 0 if $0 < \rho < 1$.
Collecting our results we have the following theorem.

THEOREM 6.2.13. *Let* ρ_k, *k* = 1, 2, ..., s, *where* $0 < \rho_s < \rho_{s-1} < ... < \rho_1 = 1$, *denote the s distinct real roots of the equation* $R_s(\lambda(1 - 1/\sqrt{\sigma})^2, \sigma) = 1/\sqrt{\sigma}$. *The spectral function of the M/M/s queue length process has i jumps* $(0 \le i \le s)$ *if and only if* $\rho_{i+1} \le \rho < \rho_i$, *where* $\rho_{s+1} \equiv 0$ *and* $\rho_0 \equiv \infty$. *Moreover, if* $0 < \rho < 1$ *then one jump occurs at* x = 0.

6.3   Stochastic monotonicity

In this section the results of chapter 5 will be applied to the M/M/s queue length process $\{X_i(t): 0 \le t < \infty\}$, the index $i \epsilon \{0, 1, ...\}$ indicating that

(6.3.1) $\Pr\{X_i(0) = n\} = \delta_{in}$.

As a first step we will collect all we know of $x_2 = \lim_{n\to\infty} x_{2,n}$, where $x_{2,n}$ is the second zero in ascending order of the polynomial $Q_n(x) = Q_n(x,\rho)$ defined recursively by (6.2.1). We have seen in section 2.2 that x_2 is the second point in the spectrum of the spectral function of $\{X_i(t)\}$ if the first point is isolated, and the first point in the spectrum if this is an accumulation point. The theorems 6.2.5 and 6.2.13 are readily seen to imply that there exists a critical value ρ_2 of the traffic intensity ρ, with $0 \le \rho_2 < 1$
such that

(6.3.2) $x_2 = \lambda(1 - 1/\sqrt{\rho})^2$ iff $\rho \ge \rho_2$;

if $0 < \rho < \rho_2$, then x_2 equals the smallest non-zero root of the equation

(6.3.3) $\qquad C(x)Q_{s-1}(x) = Q_s(x) ,$

where $C(x)$ is given by

(6.3.4) $\qquad C(x) = \frac{1}{2}(1 - x/\lambda + 1/\rho - \sqrt{(1 - x/\lambda + 1/\rho)^2 - 4/\rho})$

and the square root is taken positive for $x < 0$.

For purposes of numerical computation we remark that the representation (6.1.11) for $Q_n(x)$ yields after some simple algebra that

(6.3.5) $\qquad C(x)Q_{s-1}(x) - Q_s(x) = \sum_{r=0}^{s}\{C(x)\binom{s-1}{r} - \binom{s}{r}\}\prod_{i=0}^{r-1}\{(i-\rho sx/\lambda)/\rho s\} ,$

with the conventions $\binom{n}{r} = 0$ if $r > n$ and $\prod_{i=0}^{n} y_i = 1$ if $n < 0$. For completeness we mention that by theorem 6.2.13, ρ_2 is the largest root < 1 of the equation

(6.3.6) $\qquad Q_{s-1}(\lambda(1 - 1/\sqrt{\sigma})^2,\sigma)/\sqrt{\sigma} = Q_s(\lambda(1 - 1/\sqrt{\sigma})^2,\sigma) .$

The representation (6.1.11) implies that this equation is equivalent to

(6.3.7) $\qquad \sum_{r=0}^{s}\{\binom{s-1}{r}(1/\sqrt{\sigma}) - \binom{s}{r}\}\prod_{i=0}^{r-1}\{(i - \rho sx/\lambda)/\rho s\} = 0 .$

For $s = 1, 2, 3, 4, 5$ the values of ρ_2 are given in the next table.

s	ρ_2	
1	0	≈ 0.000
2	1/9	≈ 0.111
3	$2(4+\sqrt{7})/63$	≈ 0.211
4		≈ 0.284
5		≈ 0.340

Considering (6.1.4) the results of chapter 5 in terms of $\{X_i(t)\}$ may be summarized as follows.

THEOREM 6.3.1. *For the M/M/s queue length process $\{X_i(t)\}$ to be strictly stochastically increasing in the long run it is necessary and sufficient that one of the two conditions*

$\qquad$ *(i)* $\quad \rho \geq 1$

$\qquad$ *(ii)* $\quad \rho < 1$ *and* $Q_i(x_2) > 0$

be satisfied. Moreover, if (i) or (ii) holds then $\{X_i(t)\}$ is strictly stochastically increasing precisely on the interval (τ_i, ∞), where $\tau_0 = 0$ and τ_i, $i > 0$, is the unique solution of the equation

$$\int_0^\infty \exp(-xt)Q_i(x)xd\psi(x) = 0 \ .$$

In the remaining part of this section we shall show that it is computationally
simple to decide whether the process $\{X_i(t)\}$ is strictly stochastically increasing
in the long run or not. We shall assume that $\rho < 1$, which is the only interesting
case. The problem is then to decide whether $Q_i(x_2)$ is positive or not.
It was stated in theorem 6.2.13 that the spectral function of $\{X_i(t)\}$ has a jump
at $x = 0$ if $\rho < 1$. Consequently, $x_1 = 0$. We have seen earlier that $x_2 > 0$.
Considering that $x_1 = \lim_{n\to\infty} x_{1,n}$, it now follows that

$$0 = x_1 < x_{1,n} < x_2$$

for n sufficiently large. In view of the fact that the sequence $(x_{1,n})_n$ is
decreasing (see (2.2.4)) we may conclude that there exists a critical value $\tilde{n}$ of
n in the sense that

(6.3.8) $x_2 < x_{1,n}$ iff $n < \tilde{n}$,

where $\tilde{n}$ is defined to be 1 if $x_{1,n} \leq x_2$ for all n.
One readily verifies

(6.3.9) $Q_n(x) > 0$ if $0 \leq x < x_{1,n}$

and

(6.3.10) $Q_n(x) \leq 0$ if $x_{1,n} \leq x < x_{2,n}$,

where $x_{m,n} \equiv \infty$ if $m > n$. Considering that $x_2 < x_{2,n}$, we conclude

$$Q_n(x_2) > 0 \ \text{iff} \ x_2 < x_{1,n} \ .$$

Thus in view of (6.3.8) we have

(6.3.11) $Q_n(x_2) > 0$ iff $n < \tilde{n}$.

Determination of $\tilde{n}$ involves the calculation of at most $Q_1(x_2), Q_2(x_2), \ldots, Q_s(x_2)$.
Namely, if $0 < \rho < \rho_2$, then x_2 is the smallest positive root of the equation
(6.3.3) as we have seen. Considering that $R_s = Q_s/Q_{s-1}$, it appears from
figure 6.2.2, however, that this root is in the interval $(x_{1,s-1}, x_{2,s-1})$.
As a consequence, by (6.3.10), $Q_{s-1}(x_2) \lessgtr 0$, whence $\tilde{n} < s$.
If $\rho \geq \rho_2$, then, by (6.3.2),

$$x_2 = b \equiv \lambda(1 - 1/\sqrt{\rho})^2 \ .$$

From (6.1.12) and (6.1.13) we have for n > 0

$$Q_{s+n}(b) = (1/\sqrt{\rho})^n (Q_s(b)U_n(1) - Q_{s-1}(b)U_{n-1}(1)/\sqrt{\rho}).$$

It is easy to see from (6.1.14) that $U_n(1) = n+1$, whence for n > 0

(6.3.12) $$Q_{s+n}(b) = (1/\sqrt{\rho})^n (Q_s(b) - n\{Q_{s-1}(b)/\sqrt{\rho} - Q_s(b)\}).$$

Now suppose $Q_s(b) > 0$, whence $\hat{n} > s$. Considering that $\hat{n}$ is finite, i.e., $Q_{s+n}(b) < 0$ for n sufficiently large, it is implied by (6.3.12) that

$$Q_{s-1}(b)/\sqrt{\rho} - Q_s(b) > 0.$$

Hence

$$\hat{n} = \min\{n | Q_n(b) \le 0\} = s + \min\{n \mid Q_{s+n}(b) \le 0\} =$$

$$= s + \min\{n | n \ge Q_s(b)/(Q_{s-1}(b)/\sqrt{\rho} - Q_s(b))\}.$$

Summarizing our results we have the following.

THEOREM 6.3.2. *The birth-death polynomials* $Q_n(x)$, *n = 0, 1, ...,* *of the M/M/s queue length process have the property that there exists a natural number* $\hat{n}$ *such that*

$$Q_n(x_2) > 0 \ iff \ n < \hat{n}.$$

If $0 < \rho < \rho_2$, *then* $\hat{n} < s$. *If* $\rho \ge \rho_2$ *then either* $\hat{n} \le s$ *or*

$$\hat{n} = s + \min\{n | \ n \ge Q_s(b)/(Q_{s-1}(b)/\sqrt{\rho} - Q_s(b))\},$$

where $b = \lambda(1 - 1/\sqrt{\rho})^2$.

As an example we will now consider the M/M/1 queue length process. We assume the process to be positive recurrent, i.e., $\rho = \lambda/\mu < 1$. Since s = 1, one has $\rho_2 = 0$ by theorem 6.2.13. It follows that

$$x_2 = b \equiv \lambda(1 - 1/\sqrt{\rho})^2$$

for all ρ. One readily verifies from (6.2.1)

(6.3.13) $$Q_0(b) = 1, \ Q_1(b) = (2\sqrt{\rho} - 1)/\rho .$$

Consequently, by theorem 6.3.2, $\hat{n} = 1$ if $Q_1(b) \le 0$, i.e., $\rho \le \frac{1}{4}$; if $\frac{1}{4} < \rho < 1$, then

$$\hat{n} = 1 + \min\{n \mid n \ge Q_1(b)/(Q_0(b)/\sqrt{\rho} - Q_1(b))\} =$$

$$= \min\{n \mid n \ge \sqrt{\rho}/(1 - \sqrt{\rho})\} .$$

Thus, as is easy to see, we have in fact for every $\rho < 1$

(6.3.14) $\hat{n} = \min\{n \mid n \geq \sqrt{\rho}/(1 - \sqrt{\rho})\}$.

We conclude that the M/M/1 queue length process starting in state i (i.e., with i customers) at $t = 0$ is strictly stochastically increasing in the long run iff

(6.3.15) $i < \sqrt{\rho}/(1 - \sqrt{\rho})$.

In the simple case of the M/M/1 queue length process it is possible to say something about the precise interval on which the process is strictly stochastically increasing. Namely, we have established in remark 5.1.4 that the process with initial state $i > 0$ is strictly stochastically increasing precisely on the interval (τ_i, ∞), where τ_i is the solution of the equation

(6.3.16) $e_{i0}(t) = 0$.

Considering that $e_{i0}(t) = p'_{i0}(t)$ by definition of $e_{i0}(.)$, we can use the well known formula for $p'_{ij}(t)$ to solve equation (6.3.16). From, e.g., LEDERMANN and REUTER (1954), formula (4.20), or BAILEY (1954), formula (18), one finds

(6.3.17) $p'_{i0}(t) = (\sqrt{\mu/\lambda})^i \exp(-(\lambda+\mu)t)\{\sqrt{\lambda\mu}I_{i-1} - \lambda I_i - \sqrt{\lambda\mu}I_{i+1} + \lambda I_{i+2}\}$,

where $I_m, m = 0, 1, \ldots.$, are the modified Bessel functions of the first kind, the suppressed argument being $2\sqrt{\lambda\mu}t$. Since

$$I_{m-1}(t) - I_{m+1}(t) = 2mI_m(t)/t$$

(see ERDÉLYI (1953)), it follows that

$$p'_{i0}(t) = \exp(-\lambda(1+1/\rho)t)(iI_i - (i+1)\sqrt{\rho}I_{i+1})/(\sqrt{\rho})^i ,$$

where we have substituted $\mu = \lambda/\rho$. Hence the solution τ_i of equation (6.3.16), if it exists, is given by

(6.3.18) $\tau_i = t_i/2\sqrt{\lambda\mu} = t_i\sqrt{\rho}/2\lambda$,

where t_i is the solution of the equation

(6.3.19) $I_i(t)/I_{i+1}(t) = (1 + 1/i)\sqrt{\rho}$ $t > 0$.

It is interesting to note that $I_i(t) > I_{i+1}(t)$ for all $t > 0$, and $I_i(t)/I_{i+1}(t) + 1$ as $t \to \infty$ (see ERDÉLYI (1953)). Hence, for (6.3.19) to have a solution it is necessary that $(1 + 1/i)\sqrt{\rho} > 1$, i.e., either $\rho \geq 1$, or $\rho < 1$ and $i < \sqrt{\rho}/(1 - \sqrt{\rho})$. In view of the foregoing results this condition is also sufficient for (6.3.19) to have a unique solution.

Finally, we remark that the equation (6.3.19) occurs in the work of STANGE (1964) on the M/M/1 queue, when he looks for the point at which $P_{0i}'(t) = 0$, if it exists. The answer, of course, is τ_i since $P_{0i}(t) = \pi_i P_{i0}(t)$ by (2.1.6).

6.4   Exponential ergodicity

We recall from section 2.3 that a natural birth-death process is exponentially ergodic if the spectrum of the process has no point of accumulation at 0; the decay parameter $\hat{\alpha}_{00}$ is then equal to the first non-zero point in the spectrum. Equivalently, $\hat{\alpha}_{00} = x_1$ if $x_1 > 0$, and $\hat{\alpha}_{00} = x_2$ if $x_2 > x_1 = 0$. The results of section 6.2 for the M/M/s queue length process imply

$$x_1 = 0 \qquad \text{if } \rho \leq 1$$
$$= (1 - 1/\sqrt{\rho})^2 \quad \text{if } \rho > 1 .$$

Furthermore, there exists a value $\rho_2 < 1$ such that

$$0 < x_2 < \lambda (1 - 1/\sqrt{\rho})^2 \quad \text{if } \rho < \rho_2$$
$$x_2 = \lambda(1 - 1/\sqrt{\rho})^2 \quad \text{if } \rho \geq \rho_2 .$$

It follows that the M/M/s queue length process is exponentially ergodic iff $\rho \neq 1$. For all $\rho \neq 1$ the decay parameter $\hat{\alpha}_{00}$ is given by

(6.4.1) $\hat{\alpha}_{00} = x_2$.

We refer to section 6.3 for some computational aspects of (6.4.1)

REMARK 6.4.1. DE SMIT (1972) conjectured the existence of a critical value $\hat{\rho}$ of ρ in the sense that $\hat{\alpha}_{00} = \lambda(1-1/\sqrt{\rho})^2$ if $\rho \geq \hat{\rho}$ and $\rho \neq 1$, and $\hat{\alpha}_{00} < \lambda(1 - 1/\sqrt{\rho})^2$ if $\rho < \hat{\rho}$. Evidently, this conjecture is correct and $\hat{\rho} = \rho_2$.

REMARK 6.4.2. It is not difficult to verify that for $0 < \rho < 1$ the spectral representation of $p_{ij}(t)$ can be written as

(6.4.2) $$p_{ij}(t) = \pi_j/\Sigma\pi_n + \pi_j \int_{x_2}^{b_2} \exp(-xt)Q_i(x)Q_j(x)\,d\psi(x) ,$$

where b_2 is given in (6.2.7). As noted in theorem 2.3.1 it is possible that $\hat{\alpha}_{ij}$ is larger than $\hat{\alpha}_{00}$, to wit if $Q_i(x_2) = 0$ (or $Q_j(x_2) = 0$) and x_2 is an isolated point of the spectrum of ψ. The latter occurs iff $\rho < \rho_2$. Thus it is seen that $\hat{\alpha}_{ij} = \hat{\alpha}_{00}$ for all i, j if $\rho \geq \rho_2$ and $\rho \neq 1$.

7   A QUEUEING MODEL WHERE POTENTIAL CUSTOMERS ARE DISCOURAGED BY QUEUE
     LENGTH

7.1.   Introduction

We consider the birth-death process $\{X(t): 0 \leq t < \infty\}$ with parameters

$$(7.1.1) \quad \begin{array}{ll} \lambda_n = \lambda/(n+1) & n = 0, 1, \ldots \\ \mu_n = 0 & n = 0 \\ \quad = \mu & n = 1, 2, \ldots, \end{array}$$

which serves as a single server queueing model where potential customers are
discouraged by queue length (cf. CONOLLY (1975), HADIDI (1975) and
NATVIG (1974, 1975)).
Evidently, the potential coefficients of $\{X(t)\}$ are given by

$$(7.1.2) \quad \pi_n = (\lambda/\mu)^n/n!$$

Seeing that

$$\sum_{n=0}^{\infty}(1/\lambda_n\pi_n)\sum_{i=0}^{n}\pi_i \geq \sum_{n=0}^{\infty}1/\lambda_n = \infty$$

and

$$\sum_{n=0}^{\infty}(1/\lambda_n\pi_n)\sum_{i=n+1}^{\infty}\pi_i \geq \sum_{n=0}^{\infty}1/\mu_{n+1} = \infty,$$

we conclude that $\{X(t)\}$ is a natural birth-death process. Furthermore, we have

$$(7.1.3) \quad \Sigma \pi_n = \exp(\lambda/\mu) < \infty,$$

whence the process is positive recurrent for all positive values of λ and μ.
From (3.3.5) and (7.1.3) we obtain the steady-state results

$$(7.1.4) \quad \lim_{t\to\infty} p_{ij}(t) \equiv p_j = (\lambda/\mu)^j\exp(-\lambda/\mu)/j!.$$

To obtain the transition probabilities $p_{ij}(t)$ for finite t we let

$$(7.1.5) \quad p_{00}^*(z) = \int_0^{\infty} \exp(zt)p_{00}(t)dt \qquad \text{Re } z < 0,$$

the Laplace transform of $p_{00}(t)$. NATVIG (1974), formula (2.18), derived an
expression for $p_{i0}^*(z)$, the Laplace transform of $p_{i0}(t)$. In particular his result
implies

$$(7.1.6) \quad p_{00}^*(z) = -\frac{1}{z}\{1 + ((z-\mu)^3/\mu)\exp(-\lambda\mu/(z-\mu)^2)\sum_{k=1}^{\infty}\frac{1}{k!}\frac{(\lambda\mu/(z-\mu)^2)^k}{(z-\mu)^2-\lambda z/k}\}.$$

On the other hand, substitution of the spectral representation

$$p_{00}(t) = \int_0^\infty \exp(-xt)\,d\psi(x)$$

in (7.1.5) yields

(7.1.7) $p_{00}^*(z) = B(z) \equiv \int_0^\infty d\psi(x)/(x-z)$ Re z < 0.

Thus by (7.1.6) and (7.1.7) we have obtained an explicit expression for the Stieltjes transform of the spectral function ψ in the region Re z < 0. In the next section we shall deal with the problems of continuing $p_{00}^*(z)$ into the complex plane and of applying the inversion formula (2.2.7).

7.2   The spectral representation

We define

(7.2.1) $H(z) = -\dfrac{1}{z}\{1 + ((z-\mu)^3/\mu)\exp(-\lambda\mu/(z-\mu)^2)\sum_{k=1}^{\infty}\dfrac{1}{k!}\dfrac{(\lambda\mu/(z-\mu)^2)^k}{(z-\mu)^2-\lambda z/k}\}$.

Clearly, H(z) has an essential singularity at the point μ and a simple pole at 0. Furthermore, H(z) has singularities at the zeros of $(z-\mu)^2 - \lambda z/k$, for k = 1,2,... . We have

(7.2.2) $(z-\mu)^2 - \lambda z/k = (z - a_k)(z - b_k)$,

where a_k and b_k are given by

(7.2.3) $a_k = \mu - (\sqrt{\lambda^2 + 4k\lambda\mu} - \lambda)/2k$

and

(7.2.4) $b_k = \mu + (\sqrt{\lambda^2 + 4k\lambda\mu} + \lambda)/2k$.

One readily verifies

(7.2.5) $0 < a_k < a_{k+1} < \mu < b_{k+1} < b_k < \lambda + 2\mu$.

Moreover

(7.2.6) $a_k \to \mu,\ b_k \to \mu$ as k → ∞ .

Considering the above we see that H(z) has simple poles at the points 0, a_k and b_k, k = 1, 2, ..., and no other singularities besides these and the point μ. After some simple algebra one gets the following results for the residues of the poles of -H(z).

(7.2.7) $\text{Res}_0 \ -H(z) = \exp(-\lambda/\mu)$

(7.2.8) $\text{Res}_{a_k} \ -H(z) = \dfrac{\lambda(\mu - a_k)(kb_k/\mu)^k}{\mu(b_k - a_k) \ k \ k!} \exp(-kb_k/\mu)$

(7.2.9) $\text{Res}_{b_k} \ -H(z) = \dfrac{\lambda(b_k - \mu)(ka_k/\mu)^k}{\mu(b_k - a_k) \ k \ k!} \exp(-ka_k/\mu)$.

From (7.1.6), (7.1.7) and (7.2.1) it is seen that

(7.2.10) $B(z) = H(z)$ $\hspace{5cm}$ $\text{Re } z < 0.$

Consequently $H(z)$ is the continuation of $B(z)$ into the entire complex plane with
the exception of the points 0, μ, a_k and b_k, $k = 1, 2, \ldots$. Theorem 2.2.8 now
implies that the spectral function $\psi(x)$ has a jump of magnitude
$\Delta\psi(x) = \text{Res}_x \ -H(z)$ at the point x if x is one of the points 0, a_k and b_k, $k =$
1, 2, ..., and that $\psi(x)$ is constant between these points. There remains only
one point to consider, namely μ. However, considering the previous results and the
fact that $\psi(0) = 0$ and $\psi(\infty) = 1$, we simply have

(7.2.11) $\Delta\psi(\mu) = 1 - \Delta\psi(0) - \sum\limits_{k=1}^{\infty}(\Delta\psi(a_k) + \Delta\psi(b_k))$.

We have not succeeded in evaluating the expression (7.2.11). It is conjectured,
however, that

(7.2.12) $\Delta\psi(\mu) = 0$.

In view of the preceding results regarding $\psi(x)$ we must find explicit
expressions for $Q_n(0)$, $Q_n(\mu)$, $Q_n(a_k)$ and $Q_n(b_k)$, $k = 1, 2, \ldots$, in order to
obtain an explicit expression for the representation formula

(7.2.12) $p_{ij}(t) = \pi_j \int\limits_0^{\infty} \exp(-xt)Q_i(x)Q_j(x) \, d\psi(x)$.

Now, the recurrence formulas (2.2.1) for the polynomials $Q_n(x)$ can be written
in the pertinent case as

(7.2.13)
$$Q_0(x) = 1$$
$$Q_1(x) = 1 - x/\lambda$$
$$Q_n(x) = (1 + n\mu/\lambda - nx/\lambda)Q_{n-1}(x) - (n\mu/\lambda)Q_{n-2}(x) \hspace{2cm} n > 1.$$

From these relations one easily gets

(7.2.14) $Q_n(0) = 1$ $\hspace{5cm}$ $n = 0, 1, \ldots$.

To obtain $Q_n(x)$ for $x = \mu$, a_k and b_k, $k = 1, 2, \ldots$, we introduce the functions

$$c_0(x) = 0$$

(7.2.15)

$$c_n(x) = ((\lambda/\mu)^n/n!)(Q_{n-1}(\mu x) - Q_n(\mu x))/x \qquad n > 0.$$

It is not difficult to see with (7.2.13) that

(7.2.16) $\qquad Q_n(\mu x) = (n!/(\lambda/\mu)^n)(c_{n+1}(x) - c_n(x)) \qquad n \geq 0.$

In fact one has $c_n(x) = Q_{n-1}^*(\mu x)$, where $\{Q_n^*(x)\}_n$ are the dual polynomials of $\{Q_n(x)\}_n$ as defined in chapter 3; hence (7.2.16) follows at once from theorem 3.2.2. The functions $c_n(x)$ are also easily seen to satisfy the recurrence relations

$$c_0(x) = 0 \ , \ c_1(x) = 1$$

(7.2.17)

$$nc_{n+1}(x) = (\lambda/\mu + n(1-x))c_n(x) - (\lambda/\mu)c_{n-1}(x) \qquad n > 0.$$

With

(7.2.18) $\qquad C(x,y) = \sum_{n=0}^{\infty} c_{n+1}(x)y^n \ , \ C'(x,y) = \sum_{n=1}^{\infty} nc_{n+1}(x)y^{n-1}$

one finds from (7.2.17)

(7.2.19) $\qquad (1 + y(x-1))C'(x,y) = (\lambda(1-y)/\mu + 1 - x)C(x,y) \ .$

In particular

$$C'(1,y) = \lambda(1-y)C(1,y)/\mu \ ,$$

so that

(7.2.20) $\qquad C(1,y) = \exp(\lambda y(1 - \tfrac{1}{2}y)/\mu).$

Hence

(7.2.21) $\qquad c_{n+1}(1) = \sum_{j=0}^{[n/2]} (-\tfrac{1}{2})^j (\lambda/\mu)^{n-j}/(n-2j)!j! \ .$

After some simple algebra it follows from (7.2.16) and (7.2.21) that

(7.2.22) $\qquad Q_n(\mu) = \sum_{j=0}^{[(n+1)/2]} \binom{n+1}{2j} (2j)! \ (-\mu/2\lambda)^j/j!$

For $x \neq 1$, (7.2.19) may be written as

(7.2.23) $\qquad \dfrac{C'(x,y)}{C(x,y)} = \dfrac{\lambda/\mu}{1-x} + \dfrac{\lambda x/\mu - (x-1)^2}{(x-1)^2} \left[\dfrac{x-1}{1 + (x-1)y} \right] \ ,$

whence

(7.2.24) $\qquad C(x,y) = \exp(\lambda y/\mu(1-x))(1 + (x-1)y)^{d(x)} \ ,$

where

(7.2.25) $\qquad d(x) = (\lambda x/\mu - (x-1)^2)/(x-1)^2$.

From (7.2.2) one gets

(7.2.26) $\qquad (a_k - \mu)^2 = \lambda a_k/k$,

whence

(7.2.27) $\qquad d(a_k/\mu) = k-1$.

Consequently,

(7.2.28) $\qquad C(a_k/\mu,y) = \exp(\lambda y/(\mu-a_k))(1 + (a_k-\mu)y/\mu)^{k-1}$.

Thus

(7.2.29) $\qquad c_{n+1}(a_k/\mu) = \sum\limits_{i+j=n}\binom{k-1}{j}((a_k-\mu)/\mu)^j (\lambda/(\mu-a_k))^i/i! =$

$\qquad\qquad = (\lambda/(\mu-a_k))^n \sum\limits_{j=0}^{n}\binom{k-1}{j}(-a_k/\mu k)^j/(n-j)!$.

It follows from (7.2.16) and (7.2.29) that

(7.2.30) $\qquad Q_n(a_k) = (\mu/(\mu-a_k))^n \sum\limits_{j=0}^{n}\binom{n}{j}\binom{k-1}{j}j!\{1-(n-j)(\mu-a_k)/\lambda\}(-a_k/\mu k)^j$.

Analogously, once can show

(7.2.31) $\qquad Q_n(b_k) = (\mu/(\mu-b_k))^n \sum\limits_{j=0}^{n}\binom{n}{j}\binom{k-1}{j}j!\{1+(n-j)(b_k-\mu)/\lambda\}(-b_k/\mu k)^j$.

Substitution of our results in (7.2.12) yields the following theorem.

THEOREM 7.2.1. *The transition probability* $p_{ij}(t)$ *of a birth-death process with parameters* $\lambda_n = \lambda/(n+1)$, $\mu_0 = 0$ *and* $\mu_n = \mu$ $(n > 0)$ *is given by*

$$p_{ij}(t) = \{(\lambda/\mu)^j/j!\}\{\exp(-\lambda/\mu) + \sum\limits_{k=1}^{\infty}\exp(-a_k t)Q_i(a_k)Q_j(a_k)\Delta\psi(a_k) +$$

$$+ \exp(-\mu t)Q_i(\mu)Q_j(\mu)\Delta\psi(\mu) + \sum\limits_{k=1}^{\infty}\exp(-b_k t)Q_i(b_k)Q_j(b_k)\Delta\psi(b_k)\} ,$$

where a_k *and* b_k *are given by* (7.2.3) *and* (7.2.4); $\Delta\psi(a_k) = \text{Res}_{a_k} -H(z)$ *and* $\Delta\psi(b_k) = \text{Res}_{b_k} -H(z)$ *by* (7.2.8) *and* (7.2.9); *and, finally,* $\Delta\psi(\mu)$, $Q_n(\mu)$, $Q_n(a_k)$ *and* $Q_n(b_k)$ *by* (7.2.11), (7.2.22), (7.2.30) *and* (7.2.31); *respectively.*

7.3   Stochastic monotonicity and exponential ergodicity

It is evident from (7.2.3) and (7.2.5) that the smallest point $\neq 0$ in the support
of the spectral function ψ is

$$a_1 = \mu - (\sqrt{\lambda^2+4\lambda\mu} - \lambda)/2 \ .$$

Thus we have by the results of section 2.2

$$x_2 = \mu - (\sqrt{\lambda^2+4\lambda\mu} - \lambda)/2 \ .$$

From theorem 5.3.2 we know that the process $\{X_i(t): 0 \leq t < \infty\}$, the index
$i \in \{0, 1, ...\}$ indicating that $\Pr\{X_i(0) = n\} = \delta_{in}$, is strictly stochastically
increasing in the long run iff $Q_i(x_2) > 0$. From (7.2.30) we obtain

$$Q_n(a_1) = (\mu/(\mu-a_1))^n (1 - n(\mu-a_1)/\lambda),$$

so that

$$Q_n(a_1) > 0 \quad \text{iff} \quad n < \lambda/(\mu-a_1) \ .$$

Summarizing the next theorem holds.

THEOREM 7.3.1. *The birth-death process with initial state i and parameters*
$\lambda_n = \lambda/(n+1)$, $\mu_0 = 0$ *and* $\mu_n = \mu$ $(n > 0)$ *is strictly stochastically increasing*
in the long run iff $i < (\sqrt{\lambda^2+4\lambda\mu} + \lambda)/2\mu$.

The following is evident.

THEOREM 7.3.2. *The birth-death process with parameters* $\lambda_n = \lambda/(n+1)$, $\mu_0 = 0$
and $\mu_n = \mu$ $(n > 0)$ *is exponentially ergodic; the decay parameter* $\hat{a}_{00}$ *of* $p_{00}(t)$
is given by $\hat{a}_{00} = \mu - (\sqrt{\lambda^2+4\lambda\mu} - \lambda)/2$.

8   LINEAR GROWTH BIRTH-DEATH PROCESSES

8.1   Introduction

In this chapter our subject will be birth-death processes with parameters

$$(8.1.1) \qquad \lambda_n = \alpha + \lambda n \qquad\qquad n = 0, 1, \ldots$$
$$\mu_n = \beta + \mu n \qquad\qquad n = 0, 1, \ldots \; .$$

Such processes occur naturally in the study of biological reproduction and population growth (see, e.g., GOEL and RICHTER-DYN (1974)).

CALLAERT (1971), chapter IV.2, has studied in detail the phenomenon of exponential ergodicity in the context of these processes, hence we shall not be concerned with this aspect of linear growth, birth-death processes.

Since the results of chapter 5 give rise to interesting problems only if the process has a reflecting barrier at 0, we shall assume $\beta = 0$ so that in particular $\mu_0 = 0$. The preparatory work to get to the main results regarding stochastic monotonicity of the perinent processes is much simpler than in the preceding two chapters since KARLIN and McGREGOR (1958[a], 1958[b]) have shown that the birth-death polynomials corresponding to the parameters

$$(8.1.2) \qquad \lambda_n = \alpha + \lambda n \qquad\qquad n = 0, 1, \ldots$$
$$\mu_n = \mu n \qquad\qquad n = 0, 1, \ldots,$$

with $\alpha, \mu > 0$ and $\lambda \geq 0$, are expressible as classical orthogonal polynomials for which the spectral functions are of course well known. Before we can state these results we need some terminology.

The hypergeometric function is given by

$$(8.1.3) \qquad F(a,b;\ c;\ z) = \sum_{n=0}^{\infty} (a)_n (b)_n z^n / n! (c)_n \ ,$$

where

$$(8.1.4) \qquad (x)_0 = 1, \ (x)_n = \Gamma(x+n)/\Gamma(x) \qquad\qquad x = a, b, c.$$

We define

$$(8.1.5) \qquad \phi_n(x;\ a_1,\ a_2) = F(-n,\ -x;\ a_1;\ 1-1/a_2)$$

for $a_1 > 0$ and $0 < a_2 < 1$, and set $\phi_{-1} \equiv 0$. Then $\{\phi_n(x)\}_n$ constitute a system of polynomials and

$$(8.1.6) \qquad m_n(x;\ a_1,\ a_2) = (a_1)_n \phi_n(x;\ a_1,\ a_2)$$

are the *Meixner polynomials* (see ERDÉLYI (1953)), which are orthogonal with respect to a jump function with jumps at $x = 0, 1, 2, \ldots$.

Let $L_n^\gamma(x)$ denote the nth *Laguerre polynomial* with parameter $\gamma > -1$ (see ERDÉLYI (1953)). These polynomials are orthogonal with respect to a spectral function ψ defined on the positive real axis by

(8.1.7) $\psi(x) = \exp(-x) x^\gamma$.

Finally, we recall (see (6.1.9)) that $c_n(x, a)$ denotes the nth Poisson-Charlier polynomial with parameter $a > 0$. It can be found in ERDÉLYI (1953) that the Poisson-Charlier polynomials are orthogonal with respect to a jump function with jumps at $x = 0, 1, 2, \ldots$.

The birth-death polynomials defined recursively by (2.2.1) which correspond to the parameters (8.1.2) are as follows,

(8.1.8) $\lambda = 0$: $Q_n(x) = c_n(x/\mu, \alpha/\mu)$

(8.1.9) $0 < \lambda < \mu$: $Q_n(x) = \phi_n(x/(\mu-\lambda); \alpha/\lambda, \lambda/\mu)$

(8.1.10) $\lambda = \mu$: $Q_n(x) = L_n^\gamma(x/\lambda) / \binom{n+\gamma}{n}$, $\gamma = (\alpha/\lambda) - 1$

(8.1.11) $\lambda > \mu$: $Q_n(x) = (\mu/\lambda)^n \phi_n((x/(\lambda-\mu)) - (\alpha/\lambda); \alpha/\lambda, \mu/\lambda)$.

The result (8.1.8), pertaining to the M/M/∞ queue, was given by KARLIN and McGREGOR (1958[a]); the results (8.1.9) – (8.1.11) were derived by KARLIN and McGREGOR (1958[b]).

We finally note that the potential coefficients π_n for the linear growth, birth-death process with parameters (8.1.2) are given by

(8.1.12) $\pi_n = (\alpha/\mu)^n/n!$ if $\lambda = 0$

and

(8.1.13) $\pi_n = (1+\alpha/\lambda)_n (\lambda/\mu)^n/n!$ if $\lambda > 0$.

The fact that we are dealing with natural birth-death processes is easily verified.

8.2   Stochastic monotonicity

Consider the linear growth, birth-death process $\{X_i(t): 0 \le t < \infty\}$
with initial state i and parameters $\lambda_n = \alpha + \lambda n$, $\mu_n = \mu n$. It is easily seen
from (8.1.12) and (8.1.13) that

(8.2.1) $\Sigma \, \pi_n < \infty$ iff $\lambda < \mu$.

By theorem 5.2.2 we therefore have that $\{X_i(t)\}$ is strictly stochastically
increasing in the long run if $\lambda \ge \mu$. If $\lambda < \mu$, however, then, by theorem 5.3.2,
the sign of $Q_i(x_2)$ determines whether $\{X_i(t)\}$ is strictly stochastically
increasing in the long run or not, where x_2 is the second point in the support
of the spectral function. Let us consider the case $\lambda = 0$ first.
The results of the foregoing section imply that $x_2 = \mu$ in this case, whence
by (8.1.8)

(8.2.2) $Q_n(x_2) = Q_n(\mu) = c_n(1, \alpha/\mu)$.

Poisson-Charlier polynomials have the property

$$c_n(x,a) = c_x(n,a)$$

(ERDÉLYI (1953)). Thus, in particular

$$c_n(1, \alpha/\mu) = c_1(n, \alpha/\mu) .$$

Hence, by (6.1.9),

(8.2.3) $Q_n(x_2) = c_1(n, \alpha/\mu) = 1 - n\mu/\alpha$.

It follows that

(8.2.4) $Q_n(x_2) > 0$ iff $n < \alpha/\mu$.

Next consider the case $0 < \lambda < \mu$. The results of the previous section imply
that $x_2 = \mu - \lambda$, whence by (8.1.9)

(8.2.5) $Q_n(x_2) = \phi_n(1; \alpha/\lambda, \lambda/\mu)$.

It follows by (8.1.5) and (8.1.3) that

(8.2.6) $Q_n(x_2) = 1 - n(\mu-\lambda)/\alpha$.

Hence

(8.2.7) $Q_n(x_2) > 0$ iff $n < \alpha/(\mu-\lambda)$.

It is seen from (8.2.4) that (8.2.7) is in fact valid for $0 \leq \lambda < \mu$. Thus the next theorem holds.

THEOREM 8.2.1. *For the birth-death process with initial state i and parameters $\lambda_n = \alpha + \lambda n$, $\mu_n = \mu n$ $(\alpha, \mu > 0; \lambda \geq 0)$ to be strictly stochastically increasing in the long run it is necessary and sufficient that one of the conditions (i) $\lambda \geq \mu$, (ii) $\lambda < \mu$ & $i < \alpha/(\mu-\lambda)$, be satisfied.*

9   THE MEAN OF BIRTH-DEATH PROCESSES

9.1   Introduction

Consider a natural birth-death process $\{X(t): 0 \leq t < \infty\}$ with $\mu_0 = 0$ and let $m(t)$ denote the first moment of $X(t)$, i.e.,

$$(9.1.1) \qquad m(t) = \sum_{j=0}^{\infty} j p_j(t) \ .$$

Evidently, we can also write

$$(9.1.2) \qquad m(t) = \sum_{j=0}^{\infty} \sum_{k>j} p_k(t) \ .$$

From this formulation it is readily seen that $m(t)$ is strictly monotone on (t_1, ∞) if the process $\{X(t)\}$ is strictly stochastically monotone on (t_1, ∞) (in fact, stochastic monotonicity implies the monotonicity of all moments). Thus our results on stochastic monotonicity of birth-death processes imply corresponding results on monotonicity of the mean $m(t)$. In this chapter we will investigate whether something more can be said about the behaviour of $m(t)$, given that the process starts in a fixed state. The most interesting results will be obtained in section 9.4 for a special class of birth-death processes. In section 9.2 representations will be given for $m(t)$ and its derivatives.
Section 9.3 contains some sufficient conditions for the finiteness of the mean of natural birth-death processes.

9.2   Representations

Consider a set $\{\lambda_n, \mu_n\}$ of birth-death parameters with $\mu_0 = 0$, determining (apart from the initial state distribution) a natural birth-death process $\{X(t)\}$. The parameters and variables relating to the process which is dual to $\{X(t)\}$ will as usual be indicated by an asterisk.

For $i = 0, 1, \ldots$, we define

$$(9.2.1) \qquad m_i(t) = \sum_{j=0}^{\infty} j p_{ij}(t)$$

and

$$(9.2.2) \qquad m_i^*(t) = \sum_{j=0}^{\infty} (j+1) p_{ij}^*(t) \ .$$

Further we let

$$(9.2.3) \qquad L_i(t) = \sum_{j=0}^{\infty} p_{ji}(t)$$

and for $i = -1,0,1,\ldots$

(9.2.4) $\qquad L_i^*(t) = \sum\limits_{j=0}^{\infty} p_{ji}^*(t)$.

The quantities defined by (9.2.1) – (9.2.4) may be infinite. The next lemma establishes an interesting link between the processes $\{X(t)\}$ and $\{X^*(t)\}$.

LEMMA 9.2.1.*(i)*: $m_i(t) = \sum\limits_{k=-1}^{i-1} L_k^*(t)$; *(ii)* $m_i^*(t) = \sum\limits_{k=0}^{i} L_k(t)$.

PROOF. From (3.2.3) – (3.2.4') it is easily seen that

(9.2.5) $\qquad \sum\limits_{k=N}^{\infty} p_{ik}(t) = \sum\limits_{k=-1}^{i-1} p_{N-1,k}^*(t)$

and

(9.2.6) $\qquad \sum\limits_{k=N}^{\infty} p_{ik}^*(t) = \sum\limits_{k=0}^{i} p_{Nk}(t)$.

Considering that

(9.2.7) $\qquad m_i(t) = \sum\limits_{j=0}^{\infty} \sum\limits_{k>j} p_{ik}(t)$

and

(9.2.8) $\qquad m_i^*(t) = \sum\limits_{j=0}^{\infty} \sum\limits_{k\geq j} p_{ik}^*(t)$,

the lemma follows readily. $\qquad\qquad\qquad\qquad\qquad\qquad\qquad\qquad\qquad$ □

In the remainder of this section it will be assumed that for all i and t the series (9.2.7) and (9.2.8) are convergent, or, equivalently, that for all i and t the series (9.2.3) and (9.2.4) are convergent.

Consider a finite interval $[0,T]$. Obviously, for fixed i the partial sums of the series (9.2.7) are uniformly bounded on this interval. Furthermore one has by the backward equations (1.3.5)

$$\sum\limits_{j=0}^{N} (d/dt) \sum\limits_{k=0}^{j} p_{ik}(t) = \sum\limits_{j=0}^{N} \sum\limits_{k=0}^{j} p_{ik}'(t) =$$

$$= \sum\limits_{j=0}^{N} \sum\limits_{k=0}^{j} \{\mu_i p_{i-1,k}(t) - (\lambda_i+\mu_i)p_{ik}(t) + \lambda_i p_{i+1,k}(t)\} =$$

$$= \mu_i \sum\limits_{j=0}^{N} \{1 - \sum\limits_{k>j} p_{i-1,k}(t)\} - (\lambda_i+\mu_i) \sum\limits_{j=0}^{N} \{1 - \sum\limits_{k>j} p_{ik}(t)\} + \lambda_i \sum\limits_{j=0}^{N} \{1 - \sum\limits_{k>j} p_{i+1,k}(t)\} =$$

$$= -\mu_i \sum_{j=0}^{N} \sum_{k>j} p_{i-1,k}(t) + (\lambda_i + \mu_i) \sum_{j=0}^{N} \sum_{k>j} p_{ik}(t) - \lambda_i \sum_{j=0}^{N} \sum_{k>j} p_{i+1,k}(t) \; .$$

It follows that

$$(9.2.9) \qquad \sum_{j=0}^{N} (d/dt) \sum_{k>j} p_{ik}(t) = \sum_{j=0}^{N} (d/dt)\{1 - \sum_{k=0}^{j} p_{ik}(t)\} =$$

$$= - \sum_{j=0}^{N} (d/dt) \sum_{k=0}^{j} p_{ik}(t) =$$

$$= \mu_i \sum_{j=0}^{N} \sum_{k>j} p_{i-1,k}(t) - (\lambda_i + \mu_i) \sum_{j=0}^{N} \sum_{k>j} p_{ik}(t) + \lambda_i \sum_{j=0}^{N} \sum_{k>j} p_{i+1,k}(t) \; .$$

We conclude that the series

$$(9.2.10) \qquad \sum_{j=0}^{\infty} (d/dt) \sum_{k>j} p_{ik}(t)$$

converges on $[0,\infty)$ and has uniformly bounded partial sums on $[0,T)$. Hence BENDIXSON's theorem (see BROMWICH (1965)) implies that the series (9.2.7) converges uniformly on $[0,T)$ and (9.2.9) shows that (9.2.10) also converges uniformly on every finite interval. By differentiating (9.2.9) n times and making an induction argument on n it is seen that the series

$$\sum_{j=0}^{\infty} (d^n/dt^n) \sum_{k>j} p_{ik}(t) \qquad\qquad n = 0,1,\ldots$$

all converge uniformly on every finite interval.
Similarly it can be shown that the series

$$\sum_{j=0}^{\infty} (d^n/dt^n) \sum_{k\geq j} p_{ik}^{*}(t) \qquad\qquad n = 0,1,\ldots$$

all converge uniformly on every finite interval. As a consequence $m_i(t)$ and $m_i^{*}(t)$ can be differentiated any number of times. Moreover, for $n > 0$

$$(9.2.11) \qquad m_i^{(n)}(t) \equiv (d^n/dt^n)m_i(t) = \sum_{j=0}^{\infty} (d^n/dt^n) \sum_{k>j} p_{ik}(t) =$$

$$= \sum_{j=0}^{\infty}(d^n/dt^n)\{1 - \sum_{k=0}^{j}P_{ik}(t)\} = -\sum_{j=0}^{\infty}\sum_{k=0}^{j}P_{ik}^{(n)}(t) \; ,$$

and, using (9.2.6),

$$(9.2.12) \qquad m_i^{*(n)}(t) \equiv (d^n/dt^n)m_i^*(t) = \sum_{j=0}^{\infty}(d^n/dt^n)\sum_{k \geq j}P_{ik}^*(t) =$$

$$= \sum_{j=0}^{\infty}(d^n/dt^n)\sum_{k=0}^{i}P_{jk}(t) = \sum_{j=0}^{\infty}\sum_{k=0}^{i}P_{jk}^{(n)}(t) \; ,$$

where

$$P_{ij}^{(n)}(t) = (d^n/dt^n)P_{ij}(t) \; .$$

In matrix notation the results (9.2.11) and (9.2.12) can be written compactly as

$$(9.2.13) \qquad \underline{m}^{(n)}(t) = -(P^{(n)}(t)U)\underline{1} \qquad\qquad\qquad n > 0,$$

and

$$(9.2.14) \qquad \underline{m}^{*(n)}(t) = (P^{(n)}(t)U)^T\underline{1} \qquad\qquad\qquad n > 0,$$

where

$$\underline{m}^{(n)}(t) = (m_0^{(n)}(t), m_1^{(n)}(t), \ldots.)^T \; ,$$

$$\underline{m}^{*(n)}(t) = (m_0^{*(n)}(t), m_1^{*(n)}(t), \ldots.)^T \; ,$$

and U the matrix defined by (1.2.10).
From the forward and backward equations (1.4.7) and (1.4.8) it is seen that for all n

$$P^{(n)}(t) = A^n P(t) = P(t)A^n \; .$$

Hence, for n > 0,

$$(9.2.15) \qquad P^{(n)}(t)U = (A^n P(t))U = (A^{n-1}P(t)A)U =$$

$$= A^{n-1}((P(t)A)U) = A^{n-1}E(t) \; ,$$

where E(t) is the matrix defined in (4.1.5). On the other hand it readily appears from (3.2.2) that for all n

$$A^n U = U((A^*)^T)^n \; ,$$

so that for n > 0

$$(9.2.16) \qquad P^{(n)}(t)U = (P(t)A^n)U = (P(t)A)(A^{n-1}U) =$$

$$= (P(t)A)(U((A^*)^T)^{n-1}) = ((P(t)A)U)((A^*)^T)^{n-1} = E(t)((A^*)^T)^{n-1}.$$

Substitution of the results (9.2.16) and (9.2.15) in (9.2.13) and (9.2.14), respectively, yields the next lemma, where $n \geq 0$ and

$$(9.2.17) \qquad B = (A^*)^T .$$

LEMMA 9.2.2. *(i)* $\underline{m}^{(n+1)}(t) = -(E(t)B^n)\underline{1}$; *(ii)* $\underline{m}^{*(n+1)}(t) = (A^n E(t))^T \underline{1}$.

9.3   Sufficient conditions for finiteness

Theorem 8 of KARLIN and McGREGOR (1957[a]) states that the series

$$\sum_{j=0}^{\infty} P_{ij}(t) Q_j(x) ,$$

where $\{P_{ij}(t)\}$ are the transition probabilities and $\{Q_j(x)\}$ the polynomials of a natural birth-death process with $\mu_0 \geq 0$, converges uniformly on every bounded region $0 \leq t < T$, $0 \leq |x| \leq R$. Considering that $Q_j(x) > 0$ for $x \leq 0$, it follows that for any $\hat{x} < 0$

$$\sum_{j=0}^{\infty} j P_{ij}(t) |Q_j(\hat{x})/j| = \sum_{j=0}^{\infty} P_{ij}(t) Q_j(\hat{x}) < \infty ,$$

whence

$$\sum_{j=0}^{\infty} j P_{ij}(t) < \infty \quad \text{if} \quad \liminf_{j \to \infty} Q_j(\hat{x})/j > 0 .$$

In particular

$$(9.3.1) \qquad \sum_{j=0}^{\infty} j P_{ij}(t) < \infty \quad \text{if} \quad \liminf_{j \to \infty} Q_j(-1)/j > 0 .$$

From KEMPERMAN (1962) we obtain for $x > 0$

$$(9.3.2) \qquad Q_j(-x) = \sum_{r=0}^{j} q_{jr} x^r ,$$

where

$$(9.3.3) \qquad
\begin{aligned}
q_{jr} &= 1 + \mu_0 \sum_{k=0}^{j-1} 1/\lambda_k \mu_k & \text{if} && r = 0 \\[1em]
&= \sum_{k=0}^{j-1} \pi_k q_{k,r-1} \sum_{n=k}^{j-1} 1/\lambda_n \mu_n & \text{if} && 0 < r \leq j \\[1em]
&= 0 & \text{if} && r > j .
\end{aligned}$$

It follows that

$$(9.3.4) \qquad Q_j(-1) = \sum_{r=0}^{j} q_{jr} > q_{j1} \geq \sum_{k=0}^{j-1} \pi_k \sum_{n=k}^{j-1} 1/\lambda_n \pi_n = \sum_{n=0}^{j-1} (1/\lambda_n \pi_n) \sum_{k=0}^{n} \pi_k .$$

Combining (9.3.1) and (9.3.4) yields the next theorem.

THEOREM 9.3.1. *Let* $\{\lambda_n, \mu_n\}$ *be the set of parameters of a natural birth-death process. If*

$$\liminf_{j\to\infty} (1/j)\sum_{n=0}^{j-1}(1/\lambda_n\mu_n)\sum_{k=0}^{n}\pi_k > 0,$$

then, for every $i = 0, 1, \ldots,$ *the series* $m_i(t) = \sum_{j=0}^{\infty}jP_{ij}(t)$ *converges uniformly on every finite interval.*

COROLLARY 9.3.2. *If* $\lambda_n = O(1),$ $(n\to\infty),$ *then* $\sum_{j=0}^{\infty}jP_{ij}(t)$ *converges uniformly on every finite interval.*

PROOF. If $\lambda_n < K$, say, then for all j

$$(1/j)\sum_{n=0}^{j-1}(1/\lambda_n\pi_n)\sum_{k=0}^{n}\pi_k > (1/j)\sum_{n=0}^{j-1}1/\lambda_n > 1/K > 0. \qquad \square$$

The aforementioned theorem of KARLIN and McGREGOR also leads to the following. Let $\hat{x} < 0$, then

$$\sum_{j=0}^{\infty}P_{ji}(t)|\pi_j Q_j(\hat{x})| = \pi_i\sum_{j=0}^{\infty}P_{ij}(t)Q_j(\hat{x}) < \infty ,$$

whence

(9.3.5) $\qquad \sum_{j=0}^{\infty}P_{ji}(t) < \infty$ if $\displaystyle\liminf_{j\to\infty} \pi_j Q_j(\hat{x}) > 0 .$

From (9.3.2) and (9.3.3) we get for x > 0

$$\pi_j Q_j(-x) = \sum_{r=0}^{j}\xi_{jr}x^r$$

where

$$\xi_{jr} = \pi_j(1 + \mu_0\sum_{k=0}^{j-1}1/\lambda_k\pi_k) \qquad \text{if} \qquad r = 0$$

$$= \pi_j\sum_{k=0}^{j-1}\xi_{k,r-1}\sum_{n=k}^{j-1}1/\lambda_n\pi_n \qquad \text{if} \qquad 0 < r \le j$$

$$= 0 \qquad \text{if} \qquad r > j .$$

From this we obtain

$$\pi_j Q_j(-1) = \sum_{r=0}^{j}\xi_{jr} > \xi_{j1} \ge \pi_j\sum_{k=0}^{j-1}\pi_k\sum_{n=k}^{j-1}1/\lambda_n\pi_n =$$

$$= \pi_j\sum_{n=0}^{j-1}(1/\lambda_n\pi_n)\sum_{k=0}^{n}\pi_k .$$

Hence, taking $\hat{x} = -1$ in (9.3.5), we find

(9.3.6) $\qquad \sum_{j=0}^{\infty}P_{ji}(t) < \infty$ if $\displaystyle\liminf_{j\to\infty} \pi_j\sum_{n=0}^{j-1}(1/\lambda_n\pi_n)\sum_{k=0}^{n}\pi_k > 0 .$

According to lemma 9.2.1, (9.2.3) and (9.2.4) we have in fact derived a sufficient condition for the process which is dual to the original process (starting in a fixed state) to have a finite mean for all t ≥ 0. Translation

of the sufficient condition in the parameters of this process (cf. section 3.1)
readily yields the next theorem.

THEOREM 9.3.2. *Let* $\{\lambda_n, \mu_n\}$ *(with* $\mu_0 \geq 0$*) be the set of parameters of a natural birth-death process. If*

$$\liminf_{j \to \infty} (\sum_{n=1}^{j} \pi_n \sum_{k=0}^{n-1} 1/\lambda_k \pi_k)/\lambda_j \pi_j > 0$$

then, for every $i = 0, 1, \ldots,$ *the series* $m_i(t) = \sum_{j=0}^{\infty} j p_{ij}(t)$ *converges uniformly on every finite interval.*

9.4   Behaviour of the mean in special cases

In this section we will discuss a natural birth-death process $\{X(t)\}$
with parameters $\{\lambda_n, \mu_n\}$ and $\mu_0 = 0$, for which the functions

$$m_i(t) = \sum_j j p_{ij}(t)$$

are finite for all i and t. Further we shall assume

(9.4.1) $\lambda_n + \mu_n = 0(n), \quad (n \to \infty) .$

The reason for imposing this condition is the following. According to lemma 9.2.2

(9.4.2) $m''(t) = -(E(t)B)\underline{1} .$

In forthcoming considerations, however, we shall use the relation

(9.4.3) $\underline{m}''(t) = -E(t)(B\underline{1}) ,$

to which, according to theorem A.2.2 of appendix 2, we are entitled iff for all i

(9.4.4) $\sum_{n=0}^{\infty} \sum_{k=m}^{\infty} e_{ik}(t)(B)_{kn} \to 0$ as $m \to \infty .$

Considering that

$$\sum_{n=0}^{\infty} \sum_{k=0}^{\infty} e_{ik}(t)(B)_{kn} = m_i''(t) < \infty ,$$

some simple algebra yields that (9.4.4) is valid iff

(9.4.5) $\mu_m e_{im}(t) - \lambda_m e_{i,m-1}(t) \to 0$ as $m \to \infty .$

From (4.3.2) we have $e_{0k}(t) = -\lambda_0 p_{k0}^*(t)$, and for $i > 0$,

$$e_{ik}(t) = \mu_i p_{k,i-1}^*(t) - \lambda_i p_{ki}^*(t) ,$$

where $P^*(t) = (p_{ij}^*(t))$ is the transition probability matrix of the process which is dual to $\{X(t)\}$. It can be shown that as a consequence of the sign variation diminishing property of $P(t)$ and the relation $V^T P^*(t) = (VP(t))^T$ (see theorem 3.2.1), the sequence of probabilities $(p_{kj}^*(t))_k$ is monotone in the long run. In fact, it decreases to zero since $L_j^*(t) = \sum\limits_k p_{kj}^*(t)$ converges to a finite limit (by lemma 9.2.1) and our assumption regarding the quantities $m_i(t)$. It follows that for all j

$$kp_{kj}^*(t) \to 0 \quad \text{as} \quad k \to \infty$$

Hence

$$ke_{ik}(t) \to 0 \quad \text{as} \quad k \to \infty .$$

Consequently, (9.4.5) holds for all i if $\lambda_m + \mu_m = O(m)$, $(m \to \infty)$, whence (9.4.3) is valid owing to (9.4.1). This result may be written as

(9.4.6) $\displaystyle m_i''(t) = -\sum_{j=0}^{\infty} \nu_j e_{ij}(t)$ $i = 0, 1, \dots$,

where

(9.4.7) $\nu_j = \lambda_{j+1} - \mu_{j+1} - \lambda_j + \mu_j$ $j = 0, 1, \dots$,

as can easily be verified.
We note that by lemma 9.2.2

(9.4.8) $\displaystyle m_i'(t) = -\sum_{j=0}^{\infty} e_{ij}(t)$ $i = 0, 1, \dots$.

Let us denote by M the class of natural birth-death parameters $\{\lambda_n, \mu_n\}$ with $\mu_0 = 0$ which satisfy (9.4.1) and for which $m_i(t) < \infty$ for all i and t. The results of this section will relate to a class of birth-death parameters which is smaller than M. Namely, we define

(9.4.9) $H^+ = \{\{\lambda_n, \mu_n\} \in M \mid \nu_j \geq \nu_{j+1} \geq 0 \text{ for } j = 0, 1, \dots\}$

and

(9.4.10) $H^- = \{\{\lambda_n, \mu_n\} \in M \mid \nu_j \leq \nu_{j+1} \leq 0 \text{ for } j = 0, 1, \dots\}$

and let

(9.4.11) $H = H^+ \cup H^- .$

The class H includes many sets of birth-death parameters of practical interest. For instance the sets defined by (6.1.1), (7.1.1) and (8.1.2) belong to H. We define

$$(9.4.12) \qquad \nu = \lim_{j \to \infty} \nu_j$$

if this exists, which is evidently the case if the corresponding set of birth-death parameters belongs to H.

Given (9.4.6), (9.4.8) and the existence of ν it is straightforwardly verified that we may write

$$(9.4.13) \qquad m_i''(t) = \sum_{j=0}^{\infty} (\nu_{j+1} - \nu_j) \sum_{k=0}^{j} e_{ik}(t) + \nu m_i'(t) \ .$$

At this point we recall the results (5.1.5), (5.1.6) and lemma 4.3.1 *(i)*, which say that $e_{ij}(t) < 0$ for $t > 0$ and j sufficiently large, and that the sequence $e_{i0}(t)$, $e_{i1}(t)$, ... has at most one change of sign. It follows that for i = 0, 1, ... and t > 0

$$(9.4.14) \qquad \sum_{k=0}^{j} e_{ik}(t) > 0 \text{ for all } j \text{ if } \sum_{k=0}^{\infty} e_{ik}(t) = -m_i'(t) \geq 0.$$

The next lemma follows readily from (9.4.13) and (9.4.14).

LEMMA 9.4.1. *(i)* If $\{\lambda_n, \mu_n\} \in H^+$ and $m_i'(t) \leq 0$, then $m_i''(t) \leq 0$.

$\qquad\qquad$ *(ii)* If $\{\lambda_n, \mu_n\} \in H^-$ and $m_i'(t) \leq 0$, then $m_i''(t) \geq 0$.

The main conclusions of this section can be drawn from lemma 9.4.1 through the following auxiliary lemma.

LEMMA 9.4.2. Let $f(t)$ be a differentiable function for $0 \leq t < \infty$.

(i) $\qquad$ If f is such that $f'(t) \leq 0$ if $f(t) \leq 0$, then $f(t+s) \leq f(t)$ for all $s \geq 0$ if $f(t) < 0$.

(ii) $\qquad$ If f is such that $f'(t) \geq 0$ if $f(t) \leq 0$, then $f(t+s) \geq 0$ for all $s \geq 0$ if $f(t) \geq 0$.

PROOF. *(i)* Let $f(t) < 0$ and suppose $f(t) < f(t+s)$ for some s, $0 < s < \infty$. It is no restriction to assume $f(t+s) < 0$. Let

$$v = \min\{u > 0 | f(t+u) = f(t+s)\},$$

It is easily seen that v exists, $0 < v \leq s$ and $f(t+u) < f(t+s) < 0$ for $0 \leq u < v$. By the theorem of the mean there exists a point w, $0 < w < v$, such that

$$f'(t+w) = (f(t+v) - f(t))/v = (f(t+s) - f(t))/v > 0 \ .$$

This is a contradiction since $f(t+w) < 0$.

(ii) is proven similarly. □

THEOREM 9.4.3. (i) If $\{\lambda_n, \mu_n\} \in H^+$, then $m_i'(t) \geq 0$ for all $i = 0, 1, \ldots$ and $t \geq 0$.

(ii) If $\{\lambda_n, \mu_n\} \in H^-$ and $m_i'(t) \geq 0$, then $m_i'(t+s) \geq 0$ for all $s \geq 0$.

PROOF. (i): Suppose $m_i'(t) < 0$ for some t. Then, by lemma 9.4.1 (i) and lemma 9.2.4 (i), $m_i'(t+s) \leq m_i'(t) < 0$ for all $s \geq 0$. Evidently, this contradicts that $m_i(u) \geq 0$ for all u.

(ii): Follows at once from lemma 9.4.1 (ii) and lemma 9.4.2 (ii). □

It is seen from theorem 9.4.3 (ii) that $m_i(t)$ is non-decreasing for $0 \leq t < \infty$ and $\{\lambda_n, \mu_n\} \in H^-$ iff $m_i'(0) \geq 0$. From (9.4.8) and (4.1.6) one obtains

(9.4.15) $m_i'(0) = \lambda_i - \mu_i$.

Hence the following holds.

THEOREM 9.4.4. If $\{\lambda_n, \mu_n\} \in H^-$, then $m_i(t)$ is non-decreasing for $t \geq 0$ iff $\lambda_i \geq \mu_i$.

Focussing our attention on the important class H^-, theorem 9.4.3 states that $m_i(t)$, $i = 0, 1, \ldots$, either is monotone on the interval $0 < t < \infty$ or has there exactly one local minimum. In theorem 9.4.4 a necessary and sufficient condition for $m_i(t)$ to be non-decreasing is given. It is natural now to ask for a necessary and sufficient condition for $m_i(t)$ to be non-increasing. A complete answer to this question is not known, but ad hoc methods may lead to nearly complete results. We shall illustrate this with an example: the M/M/s queue length process of chapter 6, which is a natural birth-death process with parameters $\lambda_n = \lambda$ and $\mu_n = n\mu$ if $n < s$, $s\mu$ if $n \geq s$.
This process has

(9.4.16) $\nu_n = -\mu$ $n < s$
         $= 0$ $n \geq s$,

whence $\{\lambda_n, \mu_n\} \in H^-$. Moreover, by (9.4.13),

(9.4.17) $m_i''(t) = \mu \sum_{k=0}^{s-1} e_{ik}(t)$.

From theorem 6.3.1 we see that the process with initial state i is stochastically increasing in the long run iff $\rho \geq 1$ or $\rho < 1$ & $Q_i(x_2) > 0$. Since, as we have remarked earlier, stochastic monotonicity implies monotonicity of the mean, a necessary condition for $m_i(t)$ to be non-increasing is given by

(9.4.18) $\qquad \rho < 1$ and $Q_i(x_2) \leq 0$.

Now suppose $\rho < 1$ and $Q_i(x_2) < 0$. By methods similar to those used in the proof of lemma 5.3.1 one can show that in this case for every j, $e_{ij}(t) > 0$ for t sufficiently large. Consequently, considering (9.4.17), $m_i''(t) > 0$ for t sufficiently large (the crucial thing being that $m_i''(t)$ is a finite sum). Now if $m_i'(u) \geq 0$, then, by theorem 9.4.3 *(ii)*, $m_i'(t) \geq 0$ for all $t > u$. Thus for t sufficiently large we have both $m_i'(t) \geq 0$ and $m_i''(t) > 0$, which contradicts the easily verifiable fact that $m_i(t)$ tends to a finite limit as t approaches infinity. Consequently, $\rho < 1$ and $Q_i(x_2) < 0$ imply that $m_i(t)$ is decreasing for $t \geq 0$.

It appears that $Q_i(x_2) = 0$ & $\rho < 1$ is the only case in which we cannot decide whether $m_i(t)$ is non-increasing or not.

10   THE TRUNCATED BIRTH-DEATH PROCESS

10.1   Introduction

A truncated birth-death process is a temporally homogeneous Markov process
$\{X(t): 0 \le t < \infty\}$ on a finite state space $\overline{S} = \{-1, 0, 1, \ldots, N, N+1\}$, say, with
transition probability functions

(10.1.1) $p_{ij}(t) = \Pr\{X(t+s) = j \mid X(s) = i\}$

which satisfy the conditions

(10.1.2) $P_{-1,j}(t) = \delta_{-1,j}$ $j \in \overline{S}, t \ge 0,$

(10.1.3) $P_{N+1,j}(t) = \delta_{N+1,j}$ $j \in \overline{S}, t \ge 0$

and for $i \in S = \{0, 1, \ldots, N\}$,

$$P_{i,i+1}(t) = \lambda_i t + o(t)$$
(10.1.4) $$P_{ii}(t) = 1 - (\lambda_i + \mu_i)t + o(t)$$
$$P_{i,i-1}(t) = \mu_i t + o(t)$$

as $t \to 0$, where λ_i and μ_i, $i \in S$, are non-negative constants. Throughout this
chapter we assume $\lambda_i > 0$ for $i \in S \backslash \{N\}$ and $\mu_i > 0$ for $i \in S \backslash \{0\}$.
In section 10.2 a number of known properties of the transition probabilities
$p_{ij}(t)$ are stated, among which the spectral representation of $p_{ij}(t)$ and the
strict total positivity of the matrix $P(t) = (p_{ij}(t))$, $i, j \in S = \{0, 1, \ldots, N\}$.
The strict total positivity of $P(t)$, or rather the fact that $P(t)$ is a sign
variation diminishing operator, which is a consequence of this property, is used
in section 10.3 to provide new proofs of results of ROSENLUND (1978) and KEILSON
(1971). In fact slight generalizations of these results are obtained.
In section 10.4 the truncated birth-death process $\{X(t)\}$ with general initial
state probabilities will be considered. Necessary and sufficient conditions are
derived for $\{X(t)\}$ to be stochastically monotone in the long run. Although it is
possible to obtain these results using the concept of dual processes, as we have
done in the case of a denumerable state space, we choose an entirely different
approach in which the concept of Sturmian sequences (see PERRON (1933)) is
fundamental.
Not surprisingly, it appears that the truncated process is much easier to analyse

than the process with denumerable state space.

10.2 Preliminaries

Using the conditions (10.1.2) – (10.1.4) and the Markovian nature of the
truncated birth-death process it is easy to show that the matrix $\overline{P}(t)$ =
$(p_{ij}(t))$, $i,j \in \overline{S}$ = $\{-1, 0, 1, \ldots, N, N+1\}$, must satisfy the initial condition

(10.2.1) $\overline{P}(0) = I$

and the differential equations

(10.2.2) $\overline{P}'(t) = \overline{A}\overline{P}(t)$

and

(10.2.3) $\overline{P}'(t) = \overline{P}(t)\overline{A}$,

where $\overline{A} = (a_{ij})$, i, $j \in \overline{S}$, is the matrix

$$(10.2.4) \quad \overline{A} = \begin{pmatrix} 0 & 0 & 0 & . & . & . & . \\ \mu_0 & -(\lambda_0+\mu_0) & \lambda_0 & 0 & . & . & . & . & . \\ 0 & \mu_1 & -(\lambda_1+\mu_1) & \lambda_1 & 0 & . & . & . & . \\ & . & . & . & . & . & . & . & . & . \\ & & . & . & . & . & . & . & . & . & . \\ & & & . & . & . .0 & \mu_N & -(\lambda_N+\mu_N) & \lambda_N \\ & & & & . & . & . & 0 & 0 & 0 \end{pmatrix}$$

It is well known (see e.g. KEMPERMAN (1962)) that the differential equations
(10.2.2) and (10.2.3) (both with initial condition (10.2.1)) have the same unique
solution, which is a stochastic semigroup, i.e., $\overline{P}(t)$ has the properties

(10.2.5) $\overline{P}(t+s) = \overline{P}(t)\overline{P}(s)$

(10.2.6) $(\overline{P}(t))_{ij} \geq 0$

and

(10.2.7) $\overline{P}(1)\underline{1} = \underline{1}$,

$\underline{1}$ denoting the column vector consisting of 1's. Clearly, the unique solution of
(10.2.1) and (10.2.2) satisfies the conditions (10.1.2) – (10.1.4), so that the
truncated birth-death process is a well-defined temporally homogeneous Markov
proces on the state space $\overline{S}$.

It is convenient to restrict out attention to the matrix $P(t) = (p_{ij}(t))$, $i, j \in S = \{0, 1, \ldots, N\}$, which is readily seen to be a substochastic semigroup, i.e., $P(t)$ satisfies the conditions

(10.2.8) $P(0) = I$

(10.2.9) $P(t+s) = P(t)P(s)$

(10.2.10) $(P(t))_{ij} \geq 0$

and

(10.2.11) $P(t)\underline{1} \leq \underline{1}$,

where vector is inequality is defined as in (1.2.15) and (1.2.16). Furthermore, $P(t)$ is seen to satisfy the differential equations

(10.2.12) $P'(t) = AP(t)$

and

(10.2.13) $P'(t) = P(t)A$,

where $A = (a_{ij})$, $i, j \in S = \{0, 1, \ldots, N\}$, i.e.,

(10.2.14)

$$A = \begin{pmatrix} -(\lambda_0+\mu_0) & \lambda_0 & 0 & \cdot & \cdot & \cdot & \cdot \\ \mu_1 & -(\lambda_1+\mu_1) & \lambda_1 & 0 & \cdot & \cdot & \cdot \\ \cdot & \cdot & \cdot & \cdot & \cdot & \cdot & \cdot \\ \cdot & \cdot & \cdot & \cdot & \cdot & \cdot & \cdot \\ & & \cdot & \cdot & \cdot & 0 & \mu_N & -(\lambda_N+\mu_N) \end{pmatrix} .$$

As for the transition probabilities $p_{ij}(t)$ with $\{i,j\} \cap \{-1, N+1\} \neq \emptyset$, one has (10.1.2) and (10.1.3). Moreover, one obtains from (10.2.1) and (10.2.3)

(10.2.15) $P_{i,-1}(t) = \mu_0 \int_0^t P_{i0}(\tau)d\tau$ $i = 0, 1, \ldots, N$

and

(10.2.16) $P_{i,N+1}(t) = \lambda_N \int_0^t P_{iN}(\tau)d\tau$ $i = 0, 1, \ldots, N$.

A number of authors have given the spectral representation of $p_{ij}(t)$, $i, j \in S$ (LEDERMANN and REUTER (1954), KEMPERMAN (1962), KEILSON (1964), KARLIN and McGREGOR (1965), ROSENLUND (1978)). Our notation is essentially the same as KARLIN and McGREGOR's (1965), pp. 354-355.

First we define the potential coefficients π_i, $i \in S$, of $\{X(t)\}$ as

(10.2.17) $\pi_0 = 1$; $\pi_n = \lambda_0\lambda_1\cdots\lambda_{n-1}/\mu_1\mu_2\cdots\mu_n$ $n = 1, 2, \ldots, N$.

Associated with the birth-death parameters λ_n and μ_n are also the polynomials $Q_n(x)$, $n \in S$, defined by the recurrence relations

$$Q_0(x) = 1$$

(10.2.18) $$\lambda_0(Q_1(x)-Q_0(x)) = \mu_0 Q_0(x) - x Q_0(x)$$

$$\lambda_n(Q_{n+1}(x)-Q_n(x)) = \mu_n(Q_n(x)-Q_{n-1}(x)) - x Q_n(x) \quad 0 < n < N ,$$

and the polynomial $Q_{N+1}(x)$ of degree N+1, defined by

(10.2.19) $$Q_{N+1}(x) - \lambda_N Q_N(x) = \mu_N(Q_N(x)-Q_{N-1}(x)) - x Q_N(x) .$$

KARLIN and McGREGOR (1965) have shown that $Q_{N+1}(x)$ has $N + 1$ distinct, real zeros $x_1 < x_2 < \ldots < x_{N+1}$. They give the spectral representation (the analogue of (2.2.3)) of $p_{ij}(t)$, i, $j \in S$, as

(10.2.20) $$p_{ij}(t) = \pi_j \sum_{k=1}^{N+1} \exp(-x_k t) \, Q_i(x_k) Q_j(x_k) \rho_k ,$$

where

(10.2.21) $$\rho_k = 1/\sum_{i=0}^{N} Q_i^2(x_k)\pi_i > 0 .$$

It follows by induction that for $x < 0$

(10.2.22) $$1 = Q_0(x) < Q_1(x) < \ldots < Q_N(x) .$$

As a consequence of (10.2.22) and (10.2.19) one has $Q_{N+1}(x) > \pi_N Q_N(x) \geq 0$ if $x < 0$, whence one concludes $x_1 \geq 0$. A more detailed statement is the following.

LEMMA 10.2.1. *If $\mu_0 = \lambda_N = 0$ then $x_1 = 0$; if $\mu_0 > 0$ or $\lambda_N > 0$ then $x_1 > 0$.*

PROOF. By (10.2.18) and (10.2.19) one has

(10.2.23) $$Q_0(0) = 1, \quad Q_n(0) = 1 + \mu_0 \sum_{k=0}^{n-1} 1/\lambda_k \mu_k \qquad n = 1, 2, \ldots, N$$

and

$$Q_{N+1}(0) = \lambda_N + \mu_0/\pi_N + \mu_0 \lambda_N \sum_{k=0}^{N-1} 1/\lambda_k \pi_k .$$

Considering that x_1, the smallest zero of $Q_{N+1}(x)$, is non-negative, the lemma follows at once.   □

From (10.2.20), (10.2.21), (10.2.23) and lemma 10.2.1 the stationary
probabilities $p_{ij} = \lim\limits_{t \to \infty} p_{ij}(t)$, i, j $\in$ S are obtained, viz.,

(10.2.24) $\qquad p_{ij} = p_j = \pi_j / \sum\limits_{k=0}^{N} \pi_k$ if $\mu_0 = \lambda_N = 0$

$\qquad\qquad\qquad\quad = 0 \qquad\qquad$ if $\mu_0 > 0$ or $\lambda_N > 0$.

It is found from (10.2.15), (10.2.16), (10.2.20) and lemma 10.2.1 that for
$\mu_0 > 0$ or $\lambda_N > 0$

(10.2.25) $\qquad P_{i,-1} = \lim\limits_{t \to \infty} P_{i,-1}(t) = \mu_0 \sum\limits_{k=1}^{N+1} Q_i(x_k)\rho_k/x_k \qquad\qquad\qquad$ i $\in$ S ,

and

(10.2.26) $\qquad P_{i,N+1} = \lim\limits_{t \to \infty} P_{i,N+1}(t) = \lambda_N \pi_N \sum\limits_{k=1}^{N+1} Q_i(x_k)Q_N(x_k)\rho_k/x_k \qquad\qquad$ i $\in$ S

(cf. TAN (1976)).

Because of (10.2.8) and (10.2.20) one has

(10.2.27) $\qquad \pi_j \sum\limits_{k=1}^{N+1} Q_i(x_k)Q_j(x_k)\rho_k = \delta_{ij} \qquad\qquad\qquad\qquad$ i, j $\in$ S .

This exhibits the fact that the polynomials $Q_n(x)$, n $\in$ S, are orthogonal
polynomials belonging to the mass distribution with masses ρ_k located at the
N + 1 points x_k. SZEGÖ (1959) gives in chapter III of his book a number of general
properties of orthogonal polynomials. Although these are formulated in terms of an
infinite system, it is easily verified by adapting SZEGÖ's proofs that the next
four lemmas hold for our finite system $\{Q_n(x): n \in S\}$.

LEMMA 10.2.2. *The zeros of the polynomials* $Q_n(x)$, n = 1, 2, ..., N, *are real and
and distinct. They are located in the interval* (x_1, x_{N+1}).

LEMMA 10.2.3 $\quad \sum\limits_{n=0}^{i} \pi_n Q_n^2(x) = \lambda_i \pi_i (Q_i'(x)Q_{i+1}(x) - Q_{i+1}'(x)Q_i(x)) \quad$ *if* i < N

$\qquad\qquad\qquad\qquad = \pi_N(Q_N'(x)Q_{N+1}(x) - Q_{N+1}'(x)Q_N(x)) \qquad$ *if* i = N .

Since $\sum\limits_{n=0}^{i} \pi_n Q_n^2(x) \geq \pi_0 Q_0^2(x) = 1$ for i $\in$ S, we have the following corollary.

COROLLARY 10.2.4. $Q_n'(x)Q_{n+1}(x) > Q_{n+1}'(x)Q_n(x) \qquad\qquad\qquad$ n $\in$ S .

We point out that as consequence of the above corollary $Q_n(x)$ and $Q_{n+1}(x)$ cannot have common zeros. A more detailed statement is the following.

LEMMA 10.2.5. *The zeros of* $Q_{n+1}(x)$ *are separated by the zeros of* $Q_n(x)$, $n = 1, 2, \ldots, N$.

Finally the next lemma holds.

LEMMA 10.2.6. *Between two zeros of* $Q_n(x)$, $n = 2, 3, \ldots, N$, *there is at least one zero of* $Q_{N+1}(x)$.

In section 10.4 we shall encounter the problem to determine whether the sequence $Q_0(x_k)$, $Q_1(x_k)$, $\ldots$, $Q_N(x_k)$ is monotone or not, where $k = 1$ if ($\mu_0 = 0$ & $\lambda_N > 0$) or ($\mu_0 > 0$ & $\lambda_N = 0$), and $k = 1, 2, \ldots$, N+1 if $\mu_0 = \lambda_N = 0$. When $k = 1$ the next theorem holds.

THEOREM 10.2.7 *(i)* *If* $\mu_0 = \lambda_N = 0$ *then* $Q_n(x_1) = 1$ *for all* $n \in S$.

   (ii) *If* $\mu_0 = 0$ & $\lambda_N > 0$ *then* $1 = Q_0(x_1) > Q_1(x_1) > \ldots > Q_N(x_1) > 0$.

   (iii) *If* $\mu_0 > 0$ & $\lambda_N = 0$ *then* $Q_N(x_1) > Q_{N-1}(x_1) > \ldots > Q_0(x_1) > 0$.

PROOF. Considering that $x_1 = 0$ if $\mu_0 = \lambda_N = 0$ by lemma 10.2.1, *(i)* follows at once from (10.2.23).
To prove *(ii)* and *(iii)* we observe that $x_1 Q_n(x_1) > 0$ if $\mu_0 > 0$ or $\lambda_N > 0$ in view of (10.2.23) and the lemmas 10.2.1 and 10.2.2. Consequently, by (10.2.18),

$$Q_1(x_1) < Q_0(x_1) = 1 \quad \text{if} \quad \mu_0 = 0 \quad \text{and} \quad \lambda_N > 0$$

and *(ii)* follows by induction. Furthermore, by (10.2.19),

$$Q_N(x_1) > Q_{N-1}(x_1) > 0 \text{ if } \mu_0 > 0 \text{ and } \lambda_N = 0$$

and *(iii)* follows by induction. □

Let $\underline{u} = (u_0, u_1, \ldots, u_m)^T$ be a vector of real numbers. As in section 2.1 we denote by $S^-(\underline{u})$ the number of sign changes in the sequence $u_0, u_1, \ldots, u_m$ by deleting all zero terms, with the special convention $S^-(\underline{0}) = -1$, $\underline{0}$ denoting the column vector consisting of zeros. The solution of the aforementioned problem when $\mu_0 = \lambda_N = 0$ is now given by the next theorem.

THEOREM 10.2.8. *Let* $\mu_0 = \lambda_N = 0$, $k \in S = \{0, 1, \ldots, N\}$ *and*

$$\underline{D}(x) = (Q_1(x) - Q_0(x), Q_2(x) - Q_1(x), \ldots, Q_N(x) - Q_{N-1}(x))^T.$$

Then $S^-(\underline{D}(x_{k+1})) = k - 1$. *Moreover* $Q_1(x_{k+1}) < Q_0(x_{k+1})$ *if* $k > 0$.

The above theorem holds when $k = 0$ as we have seen in theorem 10.2.7
(i). The proof for $k > 0$ has been relegated to appendix 4.
An important feature of the matrix $P(t)$ is that it is strictly totally positive
for $t > 0$ (KARLIN (1968), theorem 3.3.4), which means that each subdeterminant
of $P(t)$ is strictly positive for $t > 0$. An immediate consequence of this property
is

(10.2.28) $(P(t))_{ij} > 0$ $t > 0$.

Strictly totally positive (STP) matrices of finite order (as those of infinite
order) are sign variation diminishing operators. Namely, let P be an STP matrix
of order $m + 1$ and $\underline{u} = (u_0, u_1, \ldots, u_m)^T$ a vector of real numbers.
With $S^+(\underline{u})$ denoting the maximum number of sign changes possible in the
sequence $u_0, u_1, \ldots, u_m$ by allowing each zero to be replaced by ± 1, one has

(10.2.29) $S^+(P\underline{u}) \leq S^-(\underline{u})$ if $\underline{u} \neq \underline{0}$

(see KARLIN (1968), theorem 5.1.2). By copying the first part of the proof of
theorem 5.1.5 of KARLIN (1968) one can also establish

(10.2.30) $S^-(P\underline{u}) = S^+(P\underline{u}) = S^-(\underline{u})$ implies $(P\underline{u}) \neq 0$ and sign$((P\underline{u})_0)$
             equals the sign of the first non-zero component of $\underline{u}$.

In the next section we will show that the fact that $P(t)$ is STP leads to a number
of interesting characteristics of $P(t)$ in a straightforward way. We refer to
KEILSON and KESTER (1978) for further applications of the STP property of $P(t)$.

10.3   The sign structure of $P'(t)$

In this section we are interested in the sign structure of $P'(t)$, i.e., we look
for properties of the matrix $(\text{sign}(p'_{ij}(t)))$, $i, j \in S = \{0, 1, \ldots, N\}$, where,
as usual, $\text{sign}(x) = -1, 0, 1$, according as x is negative, zero or positive.
From (10.2.20) it is immediately seen that

(10.3.1) $\pi_i p_{ij}(t) = \pi_j p_{ji}(t)$.

Consequently, the sign structure of the ith column of $P'(t)$ is the same as the
sign structure of the ith row. Let $\underline{p}_i(t) = (p_{i0}(t), p_{i1}(t), \ldots, p_{iN}(t))^T$ denote
the ith row of $P(t)$. One important property of $\underline{p}'_i(t)$ is now evident from
(10.2.20), viz.,

(10.3.2) $\text{sign}(p'_{ii}(t)) = -1$ $t \geq 0.$

Other properties are obtained as follows. We find from (10.2.12)

(10.3.3) $\underline{p}'_i(t) = P^T(t)\underline{a}_i$,

where $\underline{a}_i = (a_{i0}, a_{i1}, \ldots, a_{iN})^T$ is the ith row of A. We note that $P^T(t)$ is STP
for $t > 0$ since $P(t)$ is STP for $t > 0$. Moreover, $S^-(\underline{a}_i) = 1$ if $i = 0,N$ and 2 if
$i \neq 0,N$. Hence (10.2.29) implies, for $t > 0$,

(10.3.4) $S^+(\underline{p}'_i(t)) \leq 1$ $i = 0,N$

              ≤ 2 $i \neq 0,N.$

Furthermore, we see from (10.2.30) that $\text{sign}(p'_{i0}(t)) = \text{sign}(\text{1st non-zero}$
component of $\underline{a}_i)$ if $S^-(\underline{p}'_i(t)) = S^-(\underline{a}_i)$. Considering (10.3.2) the next lemma is
now readily verified.

LEMMA 10.3.1. *If* $t > 0$, i, $j \in S$ *and* $p'_{ij}(t) \geq 0$, *then* $p'_{ik}(t) > 0$ *for* $N \geq k > j$
if $j > i$, *and for* $0 \leq k < j$ *if* $j < i$.

As a consequence of lemma 10.3.1 and (10.3.1) we have the following theorem,
which was found by ROSENLUND (1978) for the case $\mu_0 = \lambda_N = 0$.

THEOREM 10.3.2. *If* $t > 0$, i, $j \in S$ *and* $p'_{ij}(t) \geq 0$, *then* $p'_{mn}(t)$ *for all*
pairs $(m,n) \in S \times S$ *with* $(m,n) \neq (i,j)$ *and* $(m,n) \neq (j,i)$ *such that either*
$m \leq \min\{i,j\}$ *and* $n \geq \max\{i,j\}$, *or* $m \geq \max\{i,j\}$ *and* $n \leq \min\{i,j\}$.

We next study the behaviour of sign $(p'_{ij}(t))$ as t increases. Results will be
obtained for $\{i,j\} \cap \{0,N\} \neq \emptyset$.
From (10.2.9) and (10.2.13) one has

(10.3.5) $P'(t+s) = AP(t+s) = AP(t)P(s) = P'(t)P(s)$,

whence

(10.3.6) $\underline{p}'_i(t+s) = P^T(s)\underline{p}'_i(t)$.

Supposing $t > 0$ amd $p'_{i0}(t) \leq 0$, it follows from (10.3.2) and lemma 10.3.1 that
$S^-(\underline{p}'(t)) \leq 1$ and the first non-zero component of $\underline{p}'_i(t)$ is negative. Considering
that $P^T(s)$ is STP for $s > 0$, (10.2.29) and (10.3.6) therefore imply

(10.3.7) $S^+(\underline{p}'_i(t+s)) \leq 1$ $s > 0$.

Moreover, by (10.2.30), $p'_{i0}(t+s) < 0$ if $S^-(\underline{p}'_i(t+s)) = S^+(\underline{p}'_i(t+s)) = S^-(\underline{p}'_i(t))$.
If the latter does not hold , then one must have $1 = S^-(\underline{p}'_i(t)) \neq S^-(\underline{p}'_i(t+s)) = 0$,

so that $p'_{i0}(t+s) < 0$ in this case. Namely, if $p'_{i0}(t+\hat{s}) = 0$ for some $\hat{s} > 0$, then, by (10.3.7), $p'_{ij}(t+\hat{s}) < 0$ for all $j > 0$. Consequently, for $\epsilon < \hat{s}$ and $\epsilon > 0$ sufficiently small, $p'_{ij}(t+\hat{s}-\epsilon) < 0$ for all $j > 0$. Moreover $p'_{i0}(t+\hat{s}-\epsilon) \leq 0$ as we have seen. Since, by (10.3.6), $\underline{p}'_i(t+\hat{s}) = P^T(\epsilon)\underline{p}'_i(t+\hat{s}-\epsilon)$, it follows that $p'_{i0}(t+\hat{s}) < 0$, which is a contradiction. Summarizing we have

(10.3.8) $p'_{i0}(t+s) < 0$ for all $s > 0$ if $t > 0$ & $p'_{i0}(t) \leq 0$.

A similar statement concerning $p'_{iN}(t)$ is valid, whence, in view of (10.3.1), the following theorem holds.

THEOREM 10.3.3. *Let* $t > 0$, $i,j \in S$ *and* $\{i,j\} \cap \{0,N\} \neq \phi$. *Then,* $p'_{ij}(t+s) < 0$ *for all* $s > 0$ *if* $p'_{ij}(t) \leq 0$.

The above theorem was stated earlier in somewhat less general form by KEILSON (1971), corollary 1.

It is seen from theorem 10.3.3 that two types of behaviour are possible for the function $p_{ij}(t)$ if $\{i,j\} \cap \{0,N\} \neq \phi$, viz., (I) $p_{ij}(t)$ is increasing and (II) $p_{ij}(t)$ has exactly one local maximum after which it decreases to its limit p_{ij}. When $\mu_0 > 0$ or $\lambda_0 > 0$ we have, by (10.2.24), $p_{ij} = 0$, so that only the second type of behaviour can occur. In fact, since $x_1 > 0$ by lemma 10.2.1 and $Q_n(x_1) > 0$ for all n by theorem 10.2.7, the next theorem is valid in view of the representation (10.2.20).

THEOREM 10.3.4. *If* $\mu_0 > 0$ *or* $\lambda_N > 0$, *then* $p'_{ij}(t) < 0$ *for* t *sufficiently large and* $i,j \in S$.

When $\mu_0 = \lambda_N = 0$ both types of behaviour may occur. ROSENLUND (1978) gives in proposition 4 criteria for deciding which. These criteria can be simplified, however, as follows.

THEOREM 10.3.5. *Let* $\mu_0 = \lambda_N = 0$ *and* $i \in S$. *Then*

(i) $Q_i(x_2) \leq 0$ *iff* $p'_{i0}(t)$ $(= p'_{0i}(t)/\pi_i) > 0$ *for all* $t > 0$.

(ii) $Q_i(x_2) \geq 0$ *iff* $p'_{iN}(t)$ $(= \pi_N p'_{Ni}(t)/\pi_i) > 0$ *for all* $t > 0$.

PROOF. As a consequence of (10.2.20) and lemma 10.2.1 we have

(10.3.9) $p'_{i0}(t) = -\sum_{k=2}^{N+1} x_k \exp(-x_k t) Q_i(x_k)\rho_k$

and

(10.3.10) $\quad p_{iN}'(t) = -\sum\limits_{k=2}^{N+1} x_k \exp(-x_k t) Q_i(x_k) Q_N(x_k) \rho_k$.

From (10.3.9) and theorem 10.3.3 one obtains at once

(10.3.11) $\quad p_{i0}'(t) > 0$ for all $t > 0$ if $Q_i(x_2) < 0$

and

(10.3.12) $\quad Q_i(x_2) \leq 0$ if $p_{i0}'(t) > 0$ for all $t > 0$.

The points $0 = x_1$, x_2, ..., x_{N+1} are separated by the N zeros of $Q_N(x)$, according to lemma 10.2.5. Moreover $Q_N(0) = 1$, by (10.2.33). Hence

(10.3.13) $\quad \text{sign}(Q_N(x_k)) = (-1)^{k-1}$

and in particular, $Q_N(x_2) < 0$. So, in view of (10.3.10) and theorem 10.3.3, we have

(10.3.14) $\quad p_{iN}'(t) > 0$ for all $t > 0$ if $Q_i(x_2) > 0$

and

(10.3.15) $\quad Q_i(x_2) \geq 0$ if $p_{iN}'(t) > 0$ for all $t > 0$.

Now assume $Q_i(x_2) = 0$, and let y_1, y_2, ..., y_i denote the zeros of $Q_i(x)$ in ascending order. One has $x_j < y_j$ for $j = 1, 2, ..., i$, according to lemma 10.2.5, whence $x_2 = y_1$. Consequently, by lemma 10.2.6, $x_3 < y_2$. Thus $Q_i(x_3) < 0$. Since $Q_N(x_3) > 0$ by (10.3.13), it is now easily seen from (10.3.9), (10.3.10) and theorem 10.3.3 that

(10.3.16) $\quad Q_i(x_2) = 0$ implies $p_{i0}'(t) > 0$ & $p_{iN}'(t) > 0$ for all $t > 0$.

The theorem holds as a result of (10.3.11), (10.3.12), (10.3.14) - (10.3.16) . $\square$

10.4   Stochastic monotonicity

Let $\underline{q} = (q_{-1}, q_0, \ldots, q_N, q_{N+1})^T$ be the initial distribution vector of the truncated birth-death process $\{X(t): 0 \le t < \infty\}$. Then

(10.4.1) $\underline{q} \ge \underline{0}$ and $q^T\underline{1} = 1$.

Furthermore, let $p_i(t) = \Pr\{X(t) = i \mid \underline{q}\}$, $i = -1, 0, \ldots, N+1$. With $\underline{p}(t) = (p_{-1}(t), p_0(t), \ldots, p_{N+1}(t))^T$ we then have

(10.4.2) $\underline{p}^T(t) = \underline{q}^T \bar{P}(t)$ and $\underline{p}^T(t)\underline{1} = 1$.

DEFINITION 10.4.1. The process $\{X(t)\}$ is *stochastically increasing (decreasing)* on the interval (t_1, t_2) iff for every pair τ_1, τ_2 with $0 \le t_1 \le \tau_1 < \tau_2 < t_2 \le \infty$ and for all $i = 0, 1, \ldots, N+1$

(10.4.3) $\sum_{j \ge i} p_j(\tau_2) \ge \sum_{j \ge i} p_j(\tau_1)$ $(\sum_{j \ge i} p_j(\tau_2) \le \sum_{j \ge i} p_j(\tau_1))$.

The process is *strictly stochastically increasing (decreasing)* iff strict inequality prevails in (10.4.3) for $i = 1, 2, \ldots, N$.

In what follows we shall assume $\sum_{i=0}^{N} q_i > 0$, so that, by (10.2.28), for $t > 0$

(10.4.5) $p_j(t) = \sum_{i=0}^{N} q_i p_{ij}(t) > 0$ $j \in S = \{0, 1, \ldots, N\}$.

We note that as a consequence of (10.2.15) and (10.2.16)

(10.4.6) $p_{-1}(t) = q_{-1} + \mu_0 \int_0^t p_0(\tau) d\tau$

and

(10.4.7) $p_{N+1}(t) = q_{N+1} + \lambda_N \int_0^t p_N(\tau) d\tau$.

Hence $\sum_{j \ge 0} p_j(t) = 1 - p_{-1}(t)$ is either constant (if $\mu_0 = 0$) or strictly decreasing (if $\mu_0 > 0$). Similarly, $p_{N+1}(t)$ is either constant (if $\lambda_N = 0$) or strictly increasing (if $\lambda_N > 0$). It follows that the process cannot be stochastically increasing if $\mu_0 > 0$ and stochastically decreasing if $\lambda_N > 0$. We define the vector $\underline{e}(t) = (e_{-1}(t), e_0(t), \ldots, e_{N+1}(t))^T$ for $t \ge 0$ as

(10.4.8) $\underline{e}^T(t) = (\underline{p}'(t))^T U$, where U is the upper triangular matrix with entries $u_{ij} = 1$ if $j \geq i$ and $u_{ij} = 0$ otherwise. Then U^{-1} is given bu $(U^{-1})_{ij} = 1$ if $i = j$, -1 if $j = i + 1$, 0 otherwise. From (10.2.2), (10.2.5) and (10.4.2) one obtains

$$\underline{e}^T(t+s) = \underline{q}^T \bar{P}'(t+s)U = \underline{q}^T \bar{A}\bar{P}(t+s)U = \underline{q}^T \bar{A}\bar{P}(t)\bar{P}(s)U =$$

$$= \underline{q}^T \bar{P}'(t)UU^{-1}\bar{P}(s)U = \underline{e}^T(t)U^{-1}\bar{P}(s)U \ .$$

Thus

(10.4.9) $\underline{e}(t+s) = (U^{-1}\bar{P}(s)U)^T \underline{e}(t)$.

It is readily verified that $((U^{-1})^T \bar{A}U^T)_{ij} \geq 0$ for $i \neq j$, hence, by theorem 2.1 of KEILSON and KESTER (1977)

(10.4.10) $(U^{-1}\bar{P}(s)U)_{ij} \geq 0$.

The next lemma is now easily established.

LEMMA 10.4.2 *(i)* $\{X(t)\}$ *is stochastically increasing on* (t_1,t_2) *iff* $\{X(t)\}$ *is stochastically increasing on* (t_1,∞) *iff* $\mu_0 = 0$ *and* $\underline{e}(t_1) \leq \underline{0}$.
(ii) $\{X(t)\}$ *is stochastically decreasing on* (t_1,t_2) *iff* $\{X(t)\}$ *is stochastically decreasing on* (t_1,∞) *iff* $\lambda_N = 0$ *and* $\underline{e}(t_1) \geq \underline{0}$.

The following is an application of the above lemma. From (10.2.3) and (10.4.8)

(10.4.11) $\underline{e}^T(t) = \underline{q}^T \bar{P}(t)\bar{A}U$,

whence in particular

(10.4.12) $\underline{e}^T(0) = \underline{q}^T \bar{A}U$.

Consequently, $\{X(t)\}$ is stochastically increasing (decreasing) on $(0,\infty)$ iff $\underline{q}^T \bar{A}U \leq \underline{0}$ ($\underline{q}^T \bar{A}U \geq \underline{0}$). This result is equivalent to theorem 3.4 of KEILSON and KESTER (1977).
We proceed by observing that

(10.4.13) $\underline{e}^T(t) = \underline{p}^T(t)\bar{A}U$,

as a consequence of (10.4.2) and (10.2.3). Hence

(10.4.14) $e_{-1}(t) = \mu_0 p_0(t)$

(10.4.15) $e_j(t) = \mu_{j+1}p_{j+1}(t) - \lambda_j p_j(t)$ $j = 0, 1, \ldots, N-1$

(10.4.16) $e_N(t) = -\lambda_N p_N(t)$

(10.4.17) $e_{N+1}(t) = 0$.

Substitution of the representation (10.2.20) in (10.4.15) through (10.4.5)
yields for j = 0, 1, ...N-1

(10.4.18) $e_j(t) = \lambda_j \pi_j \sum_{k=1}^{N+1} \exp(-x_k t)(Q_{j+1}(x_k)-Q_j(x_k))\sum_{i=0}^{N}q_i Q_i(x_k)$.

Evidently, the process {X(t)} is strictly stochastically increasing on (t_1,∞)
if $\mu_0 = 0$ and $e_j(t) < 0$ for all $t > t_1$ and $j = 0, 1, \ldots, N$. Also, it is strictly
stochastically decreasing on (t_1,∞) if $\lambda_N = 0$ and $e_j(t) > 0$ for all $t > t_1$ and
$j = -1, 0, \ldots, N-1$. Therefore, when $\mu_0 > 0$ or $\lambda_N > 0$, we immediately obtain the
next theorem as a result of theorem 10.2.7 and (10.4.14) – (10.4.18), considering
that the first non-zero term in the above sum becomes dominant as t grows larger.

THEOREM 10.4.3. *Let* $\sum_{i=0}^{N}q_i > 0$.

(i)   *If* $\mu_0 > 0$ *and* $\lambda_N = 0$ *then {X(t)} is strictly stochastically decreasing*
        in the long run.

(ii)   *If* $\mu_0 = 0$ *and* $\lambda_N > 0$ *then {X(t)} is strictly stochastically increasing*
        in the long run.

When $\mu_0 = \lambda_N = 0$, we have $Q_i(x_1) = 1$ for all $i = 0, 1, \ldots, N$. Hence, (10.4.18)
reduces to

(10.4.19) $e_j(t) = \lambda_j \pi_j \sum_{k=1}^{N} \exp(-x_{k+1}t)(Q_{j+1}(x_{k+1})-Q_j(x_{k+1}))\sum_{i=0}^{N}q_i Q_i(x_{k+1})$

where $j = 0, 1, \ldots, N-1$. If $\sum_{i=0}^{N}q_i Q_i(x_{k+1}) = 0$ for $k = 1, 2, \ldots, N$, then the
process is not strictly stochastically monotone since $e_j(t) \equiv 0$ for all j, whence
$p_j(t)$ is constant for all j. Now suppose that $\hat{x}$, the smallest of the x_{k+1},
$k = 1, 2, \ldots, N$, for which $\sum_{i=0}^{N}q_i Q_i(x_{k+1}) \neq 0$ exists. If $\hat{x} > x_2$, it is seen from
theorem 10.2.8 and (10.4.19) that for t sufficiently large (hence for all $t > 0$
by (10.4.9)) there will be components of $\underline{e}(t)$ with opposite sign, whence {X(t)} is
nowhere stochastically monotone. If $\hat{x} = x_2$, however, then, by theorem 10.2.8 , the
non-zero components of $\underline{e}(t)$ will have the same sign for t sufficiently large.
Specifically, we have

THEOREM 10.4.4. *Let* $\mu_0 = \lambda_N = 0$.

(i)   *{X(t)} is strictly stochastically increasing in the long run iff*
        $\sum_{i=0}^{N}q_i Q_i(x_2) > 0$.

(ii)   *{X(t)} is strictly stochastically decreasing in the long run iff*
        $\sum_{i=0}^{N}q_i Q_i(x_2) < 0$.

100

Appendix 1: <u>PROOF OF THE SIGN VARIATION DIMINISHING PROPERTY OF STRICTLY TOTALLY POSITIVE MATRICES</u>

Let $\underline{u} = (u_1, u_2, \ldots)^T$ be an infinite vector of real numbers. In KARLIN's (1968) notation $S^-(\underline{u})$ denotes the number of sign changes in the sequence $u_1, u_2, \ldots$ by deleting all zero terms, with the special convention $S^-(\underline{0}) = -1$. Furthermore, $S^+(\underline{u})$ is the maximum number of sign changes possible in the sequence $u_1, u_2, \ldots$ by allowing each zero to be replaced by ± 1. For $\underline{u} \neq \underline{0}$ one has

(A.1.1) $\qquad 0 \le S^-(\underline{u}) \le S^+(\underline{u}) \le \infty$.

In the following two theorems $P = (p_{ij})$, $i, j = 1, 2, \ldots$, is a strictly totally positive matrix, i.e., every subdeterminant of P is positive. The next theorem is a generalization of theorem 22 of KARLIN and McGREGOR (1957[a]).

THEOREM A.1.1. *Let* $\underline{u} = (u_1, u_2, \ldots)^T \neq \underline{0}$ *and* $\sum_j p_{ij} u_j$ *be convergent for all* $i = 1, 2, \ldots$, *then*

$$S^+(P\underline{u}) \le S^-(\underline{u}) .$$

PROOF. Unless $S^-(\underline{u}) = n < \infty$, which we assume to be the case, there is nothing to prove. The components of $\underline{u}$ can be divided into $n+1$ groups

$$(u_1, u_2, \ldots, u_{r_1}) (u_{r_1+1}, \ldots, u_{r_2}) \ldots (u_{r_n+1}, \ldots)$$

so that each component in the first group is, say, non-negative, each component in the second group is non-positive, and generally each component in the ith group either is zero or has sign $(-1)^{i+1}$. Furthermore, there must be at least one non-zero component in each group. We let $r_0 = 0$ and $r_{n+1} = \infty$ and form

$$y_{kh} = \sum_{j=r_{h-1}+1}^{r_h} p_{kj} |u_j| \qquad h = 1, 2, \ldots, n+1; k = 1, 2, \ldots$$

Then

(A.1.2) $\qquad (P\underline{u})_k = \sum_{h=1}^{n+1} (-1)^{h-1} y_{kh}$.

With $v_k = (P\underline{u})_k$, $k = 1, 2, \ldots$, we let

$$\underline{v}^{(m)} = (v_1, v_2, \ldots, v_m)^T \qquad m = 1, 2, \ldots$$

Moreover

$$Y^{(m)} = \begin{pmatrix} y_{11} & \cdots & \cdots & y_{1n+1} \\ \cdot & \cdots & \cdots & \cdot \\ y_{m1} & \cdots & \cdots & y_{mn+1} \end{pmatrix} .$$

Consequently, (A.1.2) implies

$$\underline{v}^{(m)} = Y^{(m)} \underline{e}^+ ,$$

where

$$\underline{e}^+ = (1, -1, 1, \ldots, (-1)^n)^T .$$

For any set of natural numbers $k_1 < k_2 < \ldots < k_{n+1}$ one has

$$\det(y_{k_i j}) = \begin{vmatrix} \sum_{h=1}^{r_1} p_{k_1 h} |u_h| & \cdots & \cdots & \sum_{h=r_n+1}^{\infty} p_{k_1 h} |u_h| \\ \cdot & \cdots & \cdots & \cdot \\ \sum_{h=1}^{r_1} p_{k_{n+1} h} |u_h| & \cdots & \sum_{h=r_n+1}^{\infty} p_{k_{n+1} h} |u_h| \end{vmatrix} =$$

$$= \sum_{h_1=1}^{r_1} \sum_{h_2=r_1+1}^{r_2} \cdots \sum_{h_{n+1}=r_n+1}^{\infty} |u_{h_1}| |u_{h_2}| \cdots |u_{h_{n+1}}| \det(p_{k_i h_j}) > 0,$$

by the nature of the construction of the groups and the fact that P is strictly totally positive. Thus all $(n+1) \times (n+1)$ minors of $Y^{(m)}$ $(m > n+1)$ are positive. According to theorem 5.1.1 of KARLIN (1968) this property implies that $S^+(\underline{x}) \leq n$ for any m-dimensional vector $\underline{x}$ such that $\underline{x} = Y^{(m)} \underline{c}$ for some $(n+1)$-dimensional vector $\underline{c} \neq \underline{0}$. In particular one has $S^+(\underline{v}^{(m)}) \leq n$ for $m > n+1$, whence one concludes

$$S^+(P\underline{u}) = \lim_{m \to \infty} S^+(\underline{v}^{(m)}) \leq n = S^-(\underline{u}) . \qquad \square$$

THEOREM A.1.2. *Let* $\underline{u} = (u_1, u_2, \ldots)^T \neq \underline{0}$ *and* $\sum_j p_{ij} u_j$ *be convergent for all* $i = 1, 2, \ldots$. *If* $S^-(P\underline{u}) = S^-(\underline{u}) < \infty$, *then* $(P\underline{u})_1 \neq 0$ *and* $\text{sign}((P\underline{u})_1)$ *equals the sign of the first non-zero component of* $\underline{u}$.

PROOF. We shall use the notation of the previous proof. It is no restriction to assume that the first non-zero component of $\underline{u}$ is positive, further suppose $S^-(\underline{u}) = n < \infty$. If $S^-(P\underline{u}) = S^+(P\underline{u}) = S^-(\underline{u})$, then, for $m > n+1$ sufficiently large, $S^-(Y^{(m)}\underline{e}^+) = S^-(\underline{v}^{(m)}) = n = S^-(\underline{e}^+)$, as can readily be seen. The result (A.1.3) is easily generalized in that each subdeterminant of $Y^{(m)}$ is strictly positive for every m. Now by copying the first part of the proof of theorem 5.1.5 of KARLIN (1968) it follows that $v_1 \neq 0$ and $\text{sign}(v_1)$ equals the sign of the first non-zero component of $\underline{e}^+$, i.e., +1. □

Appendix 2: ON PRODUCTS OF INFINITE MATRICES

The expression $|\Sigma a_i| < \infty$ will mean that the series Σa_i converges to a finite limit. The expression $AB < \infty$, with A and B denoting infinite matrices, indicates that for every i and j

$$|(AB)_{ij}| = \left|\sum_k (A)_{ik}(B)_{kj}\right| < \infty.$$

The next theorem is due to MARKOV (see KNOPP (1964)).

THEOREM A.2.1. *Let* $\left|\sum_n x_{k,n}\right| < \infty$ *for all* k, $\left|\sum_k x_{k,n}\right| < \infty$ *for all* n *and* $\left|\sum_k (\sum_n x_{k,n})\right| < \infty.$ *Then,*

$$\sum_n(\sum_k x_{k,n}) = \sum_k(\sum_n x_{k,n}) \quad iff \quad \lim_{m \to \infty} \sum_{k=0}^{\infty} (\sum_{n=m}^{\infty} x_{k,n}) = 0.$$

Now let $A = (a_{ij})$, $B = (b_{ij})$ and $C = (c_{ij})$, i, j = 0, 1, ..., be infinite matrices. As a consequence of theorem A.2.1 one has the following.

THEOREM A.2.2. *Let* $AB < \infty$ *and* $BC < \infty$.

(i) $A(BC) = (AB)C < \infty$ *iff* $A(BC) < \infty$ *and* $\lim_{m \to \infty} \sum_{k=0}^{\infty} a_{ik} \sum_{n=m}^{\infty} b_{kn}c_{nj} = 0$ *for all* i,j.

(ii) $A(BC) = (AB)C < \infty$ iff $(AB)C < \infty$ *and* $\lim_{m \to \infty} \sum_{n=0}^{\infty} c_{nj} \sum_{k=m}^{\infty} a_{ik}b_{kn} = 0$ *for all* i,j.

REMARK. It will often happen that for all j, $c_{nj} = 0$ for n sufficiently large, or for all i, $a_{ik} = 0$ for k sufficiently large. Evidently, $A(BC) = (AB)C < \infty$ in this case if $AB < \infty$ and $BC < \infty$.

Appendix 3: <u>ON THE SIGN OF CERTAIN QUANTITIES</u>

We consider a non-decreasing function F which is continuous to the left and has $F(u) = 0$ for $u \leq 0$ and $F(u) \uparrow 1$ as $u \to \infty$. It is assumed that F has an infinite number of points of increase, and, finally, that

$$\int_0^\infty u^n dF(u) < \infty \qquad\qquad n = 0, 1, \ldots .$$

By $S(F)$, the *spectrum* of F, we denote the set of points of increase of F, i.e.,

(A.3.1) $S(F) = \{u \mid F(u+\varepsilon) > F(u-\varepsilon) \text{ for all } \varepsilon > 0\}$.

For $i = i, 2, \ldots$, the points $u_i = u_i(F)$ are defined recursively as

(A.3.2) $u_1 = \inf S(F)$; $u_{i+1} = \inf S(F) \backslash \{u_k : k = 1, 2, \ldots, i\}, i \geq 1$.

If u_i is a point of accumulation of $S(F)$, then, obviously, $u_k = u_i$ for all $k > i$. It is also easy to see that

$$u_i \in S(F) \qquad\qquad i = 1, 2, \ldots .$$

One even has

(A.3.3) $F(u_i + \varepsilon) > F(u_i)$ for all $\varepsilon > 0$ $\qquad\qquad i = 1, 2, \ldots .$

Now let P be a polynomial with real coefficients and consider the function

(A.3.4) $g(t) = \int_0^\infty \exp(-ut) P(u) dF(u)$ $\qquad\qquad t \geq 0$.

We are interested in obtaining $\text{sign}(g(t))$ for large values of t. The following two theorems give the solution of this problem.

THEOREM A.3.1. *If* $P(u_i) \neq 0$ *for some natural* i, *then, with* $\hat{u} = \min\{u_i \mid P(u_i) \neq 0\}$,

$$\lim_{t \to \infty} \text{sign}(g(t)) = \text{sign}(P(\hat{u})) .$$

PROOF. It is no restriction to assume $P(\hat{u}) > 0$. Since P is continuous a $\delta > 0$ exists such that $P(u) > 0$ for $\hat{u} \leq u < \hat{u} + 2\delta$. Hence, for $t \geq 0$,

$$g(t) = \int_0^\infty \exp(-ut)P(u)dF(u) = \int_{\hat{u}}^\infty \exp(-ut)P(u)dF(u) =$$

$$= \int_{\hat{u}}^{\hat{u}+2\delta} \exp(-ut)P(u)dF(u) + \int_{\hat{u}+2\delta}^\infty \exp(-ut)P(u)dF(u) \geq$$

$$\geq \int_{\hat{u}}^{\hat{u}+\delta} \exp(-ut)P(u)dF(u) + \int_{\hat{u}+2\delta}^\infty \exp(-ut)P(u)dF(u) \geq$$

$$\geq \exp(-(\hat{u}+\delta)t)\{\int_{\hat{u}}^{\hat{u}+\delta} P(u)dF(u) + \exp(-\delta t)\int_{\hat{u}+2\delta}^\infty \exp(-(u-\hat{u}-2\delta)t)P(u)dF(u)\} \ .$$

We have

$$c \equiv \int_{\hat{u}}^{\hat{u}+\delta} P(u)dF(u) > 0 \ ,$$

in view of (A.3.3) and the positivity of $P(u)$ for $\hat{u} \leq u \leq \hat{u} + \delta$. Moreover

$$|\exp(-\delta t)\int_{\hat{u}+2\delta}^\infty \exp(-(u-\hat{u}-2\delta)t)P(u)dF(u)| \leq$$

$$\leq \exp(-\delta t)\int_{\hat{u}+2\delta}^\infty |P(u)|dF(u) < c \ ,$$

for t sufficiently large. It follows that $g(t) > 0$ for t sufficiently large. □

THEOREM A.3.2. *If* $P(u_i) = 0$ *for all natural* i *(which implies that there exists an* $n < \infty$ *such that* $u_k = u_n$ *for all* $k > n$*), then, with* $\hat{u} = \max \{u_i\}$,

$$\lim_{t\to\infty} \text{sign}(g(t)) = \lim_{\delta\downarrow 0} \text{sign}(P(\hat{u}+\delta)).$$

PROOF. $P(u)$ has finitely many zeros since it is a polynomial. Hence, if $P(u_i) = 0$ for all i, then only finitely many of the u_i are distinct. Evidently, $P(\hat{u}) = 0$, but, since P is a polynomial, a $\delta > 0$ exists such that $P(u) > 0$, say, for $\hat{u} < u < \hat{u} + 2\delta$. As in the proof of theorem A.3.1 one can show

$$g(t) \geq \exp(-(\hat{u}+\delta)t)\{\int_{\hat{u}}^{\hat{u}+\delta} P(u)dF(u) + \exp(-\delta t)\int_{\hat{u}+2\delta}^\infty \exp(-(u-\hat{u}-2\delta)t)P(u)dF(u)\}.$$

Since $\hat{u}$ is a point of accumulation of $S(F)$ and $P(u) > 0$ for $\hat{u} < u \leq \hat{u} + \delta$, we have

$$c = \int_{\hat{u}}^{\hat{u}+\delta} P(u)dF(u) > 0,$$

as in the preceding proof. Also

$$\left| \exp(-\delta t) \int_{\hat{u}+2\delta}^{\infty} \exp(-(u-\hat{u}-2\delta)t)P(u)dF(u) < c, \right.$$

for t sufficiently large. It follows that $g(t) > 0$ for t sufficiently large. □

Appendix 4: <u>PROOF OF THEOREM 10.2.8</u>

For the proof of theorem 10.2.8 for k > 0 we resort to Sturm's theorem (see PERRON (1933)). Before we can state and apply this theorem, we need some preliminaries.

LEMMA A.4.1. *Let* $\underline{u} = (u_0, u_1, \ldots, u_m)^T$ *be a vector of real numbers with the properties (i)* $m > 0$, *(ii)* $u_0 \neq 0$, *(iii)* $u_m \neq 0$ *and (iv) if* $0 < i < m$ *&* $u_i = 0$ *then* $u_{i-1} u_{i+1} < 0$. *With* $\underline{v} = (v_0, v_1, \ldots, v_m)^T$ *denoting the vector with components* $v_i = (-1)^i u_i$, *one has* $S^-(\underline{u}) + S^-(\underline{v}) = m$.

PROOF. Let X_m $(m > 0)$ be the set of vectors $\underline{u} = (u_0, u_1, \ldots, u_m)^T$ satisfying the conditions *(ii)*, *(iii)* and *(iv)*, and let $P(\underline{u})$ denote the proposition: $S^-(\underline{u}) + S^-(\underline{v}) = m$. The next four statements, where $\underline{u} \in X_m$, are readily verified:

1   if $m = 1$ then $P(\underline{u})$;

2   if $m = 2$ & $u_1 = 0$ then $P(\underline{u})$;

3   if $m > 1$ & $u_{m-1} \neq 0$ & (if $\underline{w} \in X_{m-1}$ then $P(\underline{w})$) then $P(\underline{u})$;

4   if $m > 2$ & $u_{m-1} = 0$ & (if $\underline{w} \in X_{m-2}$ then $P(\underline{w})$) then $P(\underline{u})$.

The lemma follows at once.   □

DEFINITION A.4.2. A sequence of $m + 1 \geq 2$ polynomials $P_0, P_1, \ldots, P_m$ is called a *Sturmian sequence on the interval* (a,b) iff the following four conditions are satisfied:

(i) $P_m(x) \neq 0$ for $x = a,b$;

(ii) $P_0(x) \neq 0$ for $a \leq x \leq b$;

(iii)   if $P_i(x) = 0$ & $1 \leq i < m$ & $a \leq x \leq b$ then $P_{i-1}(x)P_{i+1}(x) < 0$;

(iv)   if $P_m(x) = 0$ & $a \leq x \leq b$ then $P_{m-1}(x)P_m'(x) > 0$.

The following theorem holds true (PERRON (1933)).

THEOREM A.4.3 (Sturm's theorem). *If the sequence of polynomials* $P_0, P_1, \ldots, P_m$ *is a Sturmian sequence on the interval* (a,b), *then the number of zeros of* P_m *in the interval* (a,b) *equals* $S^-(\underline{P}(a)) - S^-(\underline{P}(b))$, *where* $\underline{P}(x) = (P_0(x), P_1(x), \ldots, P_m(x))^T$.

We define the polynomials $U_i(x)$ as

(A.4.1)
$$\begin{aligned}
U_0(x) &= 1 \\
U_{i+1}(x) &= \lambda_i \pi_i (Q_{i+1}(x) - Q_i(x)) \\
U_{N+1}(x) &= \pi_N (Q_{N+1}(x) - \lambda_N Q_N(x)).
\end{aligned}$$
     $i = 0, 1, \ldots, N-1$

Since $\lambda_i \pi_i = \mu_{i+1} \pi_{i+1}$ for $i = 0, 1, \ldots, N-1$, we obtain from (10.2.18) and (10.2.19) the relations

$$(A.4.2) \quad \begin{aligned} U_0(x) &= 1; \; U_1(x) = \mu_0 - xQ_0(x)\pi_0 \\ U_{i+1}(x) &= U_i(x) - xQ_i(x)\pi_i \end{aligned} \qquad i = 1,2,\ldots,N.$$

Let $b > a > 0$ be such that $U_{N+1}(a) \neq 0$ and $U_{N+1}(b) \neq 0$, then the next lemma is valid.

LEMMA A.4.4. *The sequence of polynomials* $U_0(x), -U_1(x), U_2(x), \ldots, (-1)^{N+1} U_{N+1}(x)$ *is a Sturmian sequence on the interval* (a,b).

PROOF. The conditions (i) and (ii) of definition A.4.2 are clearly satisfied. Suppose $U_i(\hat{x}) = 0$, where $a \leq \hat{x} \leq b$ and $1 \leq i \leq N$. If $i = 1$ we have, by (A.4.1), $Q_1(\hat{x}) = Q_0(\hat{x}) = 1$, and, by (A.4.2), $U_2(\hat{x}) = -\hat{x}Q_1(\hat{x})\pi_1$. Hence $U_0(\hat{x})U_2(\hat{x}) = U_2(\hat{x}) = -\hat{x}\pi_1 < 0$, since $\hat{x} \geq a > 0$. If $i > 1$ we have, by (A.4.1), $Q_i(\hat{x}) = Q_{i-1}(\hat{x})$, and from (A.4.2) we obtain $U_{i-1}(\hat{x}) = \hat{x}Q_{i-1}(\hat{x})\pi_{i-1}$ and $U_{i+1}(\hat{x}) = -\hat{x}Q_i(\hat{x})\pi_i$. Consequently,

$$(-1)^{i-1}U_{i-1}(\hat{x}) \times (-1)^{i+1}U_{i+1}(\hat{x}) = U_{i-1}(\hat{x})U_{i+1}(\hat{x}) = -\hat{x}^2 Q_i^2(\hat{x})\pi_i \pi_{i-1}.$$

The latter is strictly negative since $Q_i(\hat{x}) = Q_{i-1}(\hat{x})$ and we know from lemma 10.2.5 that Q_i and Q_{i-1} do not have common zeros. Thus condition (iii) is satisfied. Finally, suppose $U_{N+1}(\hat{x}) = 0$ with $a \leq \hat{x} \leq b$. From (A.4.1) we see

$$U'_{N+1}(\hat{x}) = \pi_N(Q'_{N+1}(\hat{x}) - \lambda_N Q'_N(\hat{x}))$$

and from (A.4.2) we obtain $U_N(\hat{x}) = \hat{x}Q_N(\hat{x})\pi_N$. Furthermore

$$Q'_N(\hat{x})Q_{N+1}(\hat{x}) > Q'_{N+1}(\hat{x})Q_N(\hat{x})$$

by corollary 10.4.2. Combining these results we have

$$(-1)^N U_N(\hat{x}) \times (-1)^{N+1} U'_{N+1}(\hat{x}) = -\hat{x}Q_N(\hat{x})\pi_N^2(Q'_{N+1}(\hat{x}) - \lambda_N Q'_N(\hat{x})) >$$

$$-\hat{x}Q'_N(\hat{x})\pi_N^2(Q_{N+1}(\hat{x}) - \lambda_N Q_N(\hat{x})) = -\hat{x}Q'_N(\hat{x})\pi_N U_{N+1}(\hat{x}) = 0.$$

Thus also condition (iv) is satisfied. □

As a result of the above lemma, Sturm's theorem and lemma A.4.1 the next lemma holds.

LEMMA A.4.5. *Let* $b > a > 0$ *and* $U_{N+1}(x) \neq 0$ *for* $x = a,b$. *The number of zeros of* $U_{N+1}(x)$ *in the interval* (a,b) *equals* $S^-(\underline{U}(b)) - S^-(\underline{U}(a))$, *where* $\underline{U}(x) = (U_0(x), U_1(x), \ldots\ldots\ldots, U_{N+1}(x))^T$.

We recall that we must determine the number of sign changes in the sequence $Q_1(x_k)-Q_0(x_k), Q_2(x_k)-Q_1(x_k),\ldots,Q_N(x_k)-Q_{N-1}(x_k)$, i.e., in the sequence $U_1(x_k), U_2(x_k)$, $\ldots, U_N(x_k)$, for $k > 1$. A few more steps must be taken to settle the problem.

LEMMA A.4.6. *Let* $\mu_0 = 0$, *then* $S^-(\underline{U}(a)) = 1$ *for* $a > 0$ *sufficiently small.*

PROOF. It is seen from (A.4.2) that

(A.4.3) $\quad U_0(0) = 1,\ U_{i+1}(0) = \mu_0 \qquad\qquad\qquad\qquad\qquad i = 0,1,\ldots,N.$

If $\mu_0 = 0$ it follows from (10.2.23) and corollary 10.2.4 that $Q_i'(0) > Q_{i+1}'(0)$ for $i = 0,1,\ldots,N-1$, and $\lambda_N Q_N'(0) > Q_{N+1}'(0)$. Consequently, one has with (A.4.1)

(A.4.4) $\quad U_{i+1}'(0) < 0$ if $\mu_0 = 0\ \&\ 0 \le i \le N.$

The lemma follows from (A.4.3) and (A.4.4). □

LEMMA A.4.7. *Let* $\lambda_N = 0$ *and* $0 < k \le N$, *then* $S^-(\underline{U}(x_{k+1}-\varepsilon)) = S^-(\underline{U}(x_{k+1}))$ *for* $\varepsilon > 0$ *sufficiently small.*

PROOF. If $\lambda_N = 0$, then, by (A.4.1), $U_{N+1}(x_{k+1}) = \pi_N Q_{N+1}(x_{k+1}) = 0$. It follows by lemma A.4.4 and definition A.4.2 that $U_N(x_{k+1})U_{N+1}'(x_{k+1}) < 0$, whence

(A.4.5) $\quad U_N(x_{k+1}-\varepsilon)U_{N+1}(x_{k+1}-\varepsilon) \ge 0$ for $0 \le \varepsilon < \delta_{N+1},$

say. If $0 \le m \le N$ and $U_m(x_{k+1}) = 0$, then, by lemma A.4.4 and definition A.4.2, $U_{m-1}(x_{k+1})U_{m+1}(x_{k+1}) < 0$, whence

(A.4.6) $\quad U_{m-1}(x_{k+1}-\varepsilon)U_{m+1}(x_{k+1}-\varepsilon) < 0$ for $0 \le \varepsilon < \delta_m,$

say. Finally, if $U_m(x_{k+1}) > 0\ (U_m(x_{k+1}) < 0)$, then

(A.4.7) $\quad U_m(x_{k+1}-\varepsilon) > 0\ (U_m(x_{k+1}-\varepsilon) < 0)$ for $0 \le \varepsilon < \delta_m,$

say. The statements (A.4.5) – (A.4.7) are easily seen to imply that $S^-(\underline{U}(x_{k+1}-\varepsilon)) = S^-(\underline{U}(x_{k+1}))$ for $0 < \varepsilon < \delta = \min_m \delta_m$. □

Now let $\mu_0 = \lambda_N = 0$, $0 < k \le N$ and $\varepsilon > 0$ so small that (i) $\varepsilon < x_2$, (ii) $\varepsilon < x_{k+1}-x_k$, (iii) $S^-(\underline{U}(\varepsilon)) = 1$ and (iv) $S^-(\underline{U}(x_{k+1}-\varepsilon)) = S^-(\underline{U}(x_{k+1}))$. The number of zeros of $U_{N+1}(x) = \pi_N Q_{N+1}(x)$ in the interval $(\varepsilon, x_{k+1}-\varepsilon)$ equals $k-1$, since $x_1 = 0$ by lemma 10.2.1. In view of the lemmas A.4.5 – A.4.7 we have

(A.4.8) $S^-(\underline{U}(x_{k+1})) = k$ if $0 < k \leq N$ & $\mu_0 = \lambda_N = 0$.

, We further note that $U_{N+1}(x_{k+1}) = 0$. Finally (see (A.4.2)), $U_0(x_{k+1}) = 1$ and $U_1(x_{k+1})$ $= -x_{k+1} < 0$ for $k > 0$, so that one of the k sign changes in $\underline{U}(x_{k+1})$ occurs between $U_0(x_{k+1})$ and $U_1(x_{k+1})$. These observations and (A.4.5) complete the proof of theorem 10.2.8 for $k > 0$.

111

REFERENCES

AHIEZER, N.I. and KREIN, M. (1962) *Some Questions in the Theory of Moments*. Trans-
    lations of Mathematical Monographs Vol. 2. American Mathematical Society,
    Providence
AKHIEZER(AHIEZER), N.I. (1965) *The Classical Moment Problem and Some Related Questions
    in Analysis*. Oliver & Boyd, London
BAILEY, N.T.J. (1954) A continuous time treatment of a simple queue using generating
    functions. *J. Roy. Statist. Soc. Ser. B 16*, 288-291
BEREZANSKIĬ, Ju.M. (1968) *Expansions in Eigenfunctions of Selfadjoint Operators*.
    Translations of Mathematical Monographs Vol. 17. American Mathematical
    Society, Providence
BROMWICH, T.J.I'a (1965) *An Introduction to the Theory of Infinite Series*, 2nd ed. .
    MacMillan and Co., London
CALLAERT, H. (1971) Exponentiële Ergodiciteit voor Geboorte- en Sterfteprocessen
    (in Dutch). Ph.D. thesis, University of Louvain
CALLAERT, H. (1974) On the rate of convergence in birth-and-death processes. *Bull.
    Soc. Math. Belg. t. XXVI*, 173-184
CALLAERT, H. and KEILSON, J. (1973) On exponential ergodicity and spectral structure
    for birth-death processes. *Stochastic Processes Appl. 1*, 187-235
CASE, K.M. (1974) Orthogonal polynomials from the viewpoint of scattering theory.
    J. Mathematical Phys. 15, 2166-2175
CASE, K.M. (1975) Orthogonal polynomials revisited. In: *Theory and Application of
    Special Functions* (R.A. Askey, Ed.) Publication No. 35 of the Mathematics
    Research Center, The University of Wisconsin - Madison, Academic Press,
    New York
CHUNG, K.L. (1967) *Markov Chains with Stationary Transition Probabilities*, 2nd ed. .
    Springer Verlag, Berlin
CONOLLY, B. (1975) *Lecture Notes on Queueing Systems*. Ellis Horwood Limited,
    Chichester
DALEY, D.J. (1968) Stochastically monotone Markov chains. *Z. Wahrscheinlichkeits-
    theorie verw. Geb. 10*, 305-317
DOBRUŠIN, R.L. (1952) On conditions of regularity of Markov processes which are
    stationary in time and have a denumerable set of possible states (Russian).
    Uspehi Mat. Nauk 7, 185-191
ERDÉLYI, A. (Ed.) (1953) *Higher Transcendental Functions* Vol. II. McGraw-Hill Book
    Company, New York
FELLER, W. (1959) The birth and death processes as diffusion processes. *J. Math.
    Pures Appl. 38*, 301-345
FELLER, W. (1967) *An Introduction to Probability Theory and Its Applications* Vol. I,
    3rd ed. . John Wiley & Sons, New York
FREEDMAN, D.A. (1971) *Markov Chains*. Holden-Day, San Francisco
GOEL, N.S. and RICHTER-DYN, N. (1974) *Stochastic Models in Biology*. Academic Press,
    New York
HADIDI, N. (1975) A queueing model with variable arrival rates. *Per. Math. Hung. 6*,
    39-47
KARLIN, S. (1968) *Total Positivity*. Stanford University Press, Stanford
KARLIN, S. and McGREGOR, J.L. (1957a) The differential equations of birth-and-death
    processes, and the Stieltjes moment problem. *Trans. Amer. Math. Soc. 85*,
    489-546
KARLIN, S. and McGREGOR, J.L. (1957b) The classification of birth and death processes.
    Trans. Amer. Math. Soc. 86, 366-400
KARLIN, S. and McGREGOR, J.L. (1958a) Many server queueing processes with Poisson
    input and exponential service times. *Pacific J. Math. 8*, 87-118
KARLIN, S. and McGREGOR, J.L. (1958b) Linear growth, birth and death processes.
    J. Math. Mech. 7, 643-662
KARLIN, S. and McGREGOR, J.L. (1959a) A characterization of birth and death processes.
    Proc. Nat. Acad. Sci. U.S.A. 45, 375-379
KARLIN, S. and McGREGOR, J.L. (1959b) Coincidence probabilities. *Pacific J. Math. 9*,
    1141-1164

KARLIN, S. and McGREGOR, J.L. (1965) Ehrenfest urn models. *J. Appl. Probability 2*, 352-376

KEILSON, J. (1964) A review of transient behavior in regular diffusion and birth-death processes. *J. Appl. Probability 1*, 247-266

KEILSON, J. (1971) Log-concavity and log-convexity in passage time densities of diffusion and birth-death processes. *J. Appl. Probability 8*, 391-398

KEILSON, J. and KESTER, A. (1977) Monotone matrices and monotone Markov processes. *Stochastic Processes Appl. 5*, 231-241

KEILSON, J. and KESTER, A. (1978) Unimodality preservation in Markov chains. *Stochastic Processes Appl. 7*, 179-190

KEMPERMAN, J.H.B. (1962) An analytical approach to the differential equations of the birth-and-death process. *Michigan Math. J. 9(4)*, 321-361

KENDALL, D.G. (1959) Unitary dilations of one-parameter semigroups of Markov transition operators and the corresponding integral representations for Markov processes with a countable infinity of states. *Proc. London Math. Soc. (3) 9*, 417-431

KENDALL, D.G. and REUTER, G.E.H. (1957) The calculation of the ergodic projection for Markov chains and processes with a countable infinity of states. *Acta Math. 97*, 103-144

KINGMAN, J.F.C. (1963a) The exponential decay of Markov transition probabilities. *Proc. London Math. Soc. (3) 13*, 337-358

KINGMAN, J.F.C. (1963b) Ergodic properties of continuous-time Markov processes and their discrete skeletons. *Proc. London Math. Soc. (3) 13*, 593-604

KIRSTEIN, B.M. (1976) Monotonicity and comparability of time-homogeneous Markov processes with discrete state space. *Math. Operationsforsch. u. Statist. 7*, 151-168

KNOPP, K. (1964) *Theorie und Anwendung der Unendlichen Reihen*, 5th ed. . Springer Verlag, Berlin

LEDERMANN, W. and REUTER, G.E.H. (1954) Spectral theory for the differential equations of simple birth and death processes. *Philos. Trans. Roy. Soc. London Ser. A 246*, 321-369

MAKI, D.P. (1976) On birth-death processes with rational growth rates. *Siam J. Math. Anal. 7*, 29-36

NATVIG, B. (1974) On the transient state probabilities for a queueing model where potential customers are discouraged by queue length. *J. Appl. Probability 11*, 345-354

NATVIG, B. (1975) On a queueing model where potential customers are discouraged by queue length. *Scand. J. Statist. 2*, 34-42

PERRON, O, (1933) *Algebra* Tl. II. Walter de Gruyter & Co., Berlin

REICH, E. (1957) Waiting times when queues are in tandem. *Ann. Math. Statist. 28*, 768-773

REUTER, G.E.H. (1957) Denumerable Markov processes and the associated contraction semigroups on ℓ. *Acta Math. 97*, 1-46

ROSENLUND, S.I. (1978) Transition probabilities for a truncated birth-death process. *Scand. J. Statist. 5*, 119-122

SHOHAT, J.A. and TAMARKIN, J.D. (1963) *The Problem of Moments*. Mathematical Surveys Number I, rev. ed. . American Mathematical Society, Providence

de SMIT, J.H.A. (1972) The time dependent behaviour of the queue length process in the system M/M/s. CORE discussion paper no. 7217, University of Louvain

STANGE, K. (1964) Die Anlauflösung für den einfachen exponentiellen Bedienungskanal (mit beliebig vielen Warteplätzen), der für t = 0 leer ist. *Unternehmensforschung 8*, 1-24

STONE, M.H. (1964) *Linear Transformations in Hilbert Space*. American Mathematical Society Colloquium Publications Vol. XV, 2nd ed. . American Mathematical Society, New York

STOYAN, D. (1977) *Qualitative Eigenschaften und Abschätzungen stochastischer Modelle*. R. Oldebourg Verlag, Munich

SZEGÖ, S. (1959) *Orthogonal Polynomials*. American Mathematical Society Colloquium Publications Vol. XXIII, rev. ed. . American Mathematical Society, New York

TAN, W.Y. (1976) On the absorption probabilities and absorption times of finite homogeneous birth-death processes. *Biometrics 32*, 745-752

WIDDER, D.V. (1946) *The Laplace Transform*. Princeton University Press, Princeton

NOTATION INDEX

A	8,89	G	22
$\overline{A}$	88	G^*	22
A^*	22		
$\overline{A}^*$	33	H	83
a	21	H^+	83
a_k	19,67	H^-	83
$\underline{a}_i$	94	$H(z)$	15
a_{ij}	8,89		
		I	5,88
B	80		
$B(z)$	15	L	13
b	21,62	$L_i(t)$	76
b_1	47	$L_i^*(t)$	77
b_2	47	$L_n^\gamma(x)$	73
b_k	19,67		
		M	83
$C(A)$	9	$m(t)$	76
$C(z)$	47	$m_i(t)$	76
c_n	18	$m_i^*(t)$	76
$c_n(x,a)$	45	$\underline{m}^{(n)}(t)$	79
		$m_i^{(n)}(t)$	78
D_k	19	$\underline{m}^{*(n)}(t)$	79
D_k'	20	$m_i^{*(n)}(t)$	79
$D(A)$	9	$m_n(x;a_1,a_2)$	72
$D(x)$	92		
$d(x)$	70	$\hat{n}$	62
E	3	P_t	4
$E(t)$	28	$P(t)$	9,87
$\underline{e}(t)$	29,97	$P^*(t)$	23
$\underline{e}_i(t)$	29,97	$\overline{P}(t)$	88
$\underline{e}_i(t)$	38	$\underline{P}(x)$	107
$\underline{e}^*(t)$	32	$P_n(x)$	20
$e_i^*(t)$	32	$\overline{P}^*(t)$	33
$e_{ij}(t)$	28	P_j	66,91
		$\underline{p}(t)$	4,10
$F(a,b;c;z)$	72	$p_i(t)$	1
f	22	$P_{i,-1}$	91

AUTHOR INDEX

SUBJECT INDEX